The ARRL 1989-1992 Technician Class License Manual for the Radio Amateur

Edited By

Larry D. Wolfgang, WA3VIL
Randolph L. Henderson, WI5W
Joel P. Kleinman, N1BKE

Production Staff

Jean Wilson
Mark Wilson, AA2Z
David Pingree
Dianna Roy
Steffie Nelson, KA1IFB
Jodi Morin, KA1JPA
Sue Fagan
Michelle Chrisjohn, WB1ENT
Leslie Bartoloth, KA1MJP

American Radio Relay League
Newington, CT 06111 USA

When to Use This Book

In mid-1987, a committee of Volunteer Examiner Coordinators (VECs) agreed on a three-year revision cycle for question pools from which Amateur Radio examinations are designed. Under that schedule, the pool for the Element 3A (Technician class) examination was released to VECs and publishers of Amateur Radio study materials in February 1989.

This first edition of *The ARRL Technician Class License Manual* **may be used to prepare for Technician class exams administered after November 1, 1989.** This question pool will be used for at least 3 years, until October 31, 1992.

The FCC recently released the rewritten Part 97 Amateur Radio Service Rules, however. The new rules may affect the answers to some questions. The VEC Question pool Committee is expected to release supplements to the Element 2 and Element 3A Question Pools on or before March 1, 1990. These supplements will include revisions to questions and answers affected by the new rules. The supplements are expected to be put into use on July 1, 1990. Watch *QST* for details.

Foreword

We present with pride the first edition of *The ARRL Technician Class License Manual.*

The license you'll earn after passing the Technician exam conveys operating privileges in the largest amount of spectrum of any amateur license—by far. With a Technician class license, you'll be able to operate on all the spectrum allocated to the Amateur Service above 30 MHz—including the ever-popular 2-meter band.

League staff has worked diligently to gather the material you'll need to master in order to pass the written examination for the Technician class license. Every question that you might be asked is covered, as are the correct answers. The material is presented in a format that will help you learn in a classroom setting or through self study.

In addition to *The ARRL Technician Class License Manual,* you will want to have a copy of *The FCC Rule Book,* which is also published by ARRL. The *Rule Book* covers FCC regulations, with appropriate explanations.

This manual is not just the product of ARRL staff. You can do your part too. What role can you play? First, use this manual to prepare for your exam. After you have studied the material, write your suggestions or any corrections on the Feedback Form at the back of the book, and send it to us. Your input is an important part of how we improve our study materials.

David Sumner, K1ZZ
Executive Vice President, ARRL
Newington, Connecticut

HOW TO USE THIS BOOK

To earn a Technician class Amateur Radio license, you will want to know some basic electronics theory, and the rules and regulations governing the Amateur Radio Service, as contained in Part 97 of the FCC Rules. You'll also have to be able to send and receive the international Morse code at a rate of 5 WPM. This book provides a brief description of the Amateur Radio Service, and the Technician class license in particular.

The major portion of the book is designed to guide you, step by step, through the theory that you must know to pass your amateur license exam. The material is presented in a manner that closely follows the study guide, or syllabus, printed at the end of Chapter 1. Chapter 10 contains the complete Element 3A Question Pool, used on Technician exams. It also includes the multiple-choice answers and distractors that will be used by many VECs, including the ARRL VEC, on these exams. (If you have not passed the Element 2 [Novice] written exam, you will have to pass that exam before earning your Technician class license. *Tune in the World with Ham Radio*, published by ARRL, will help you pass that exam.)

At the beginning of each chapter, you will find a list of **Key Words** that appear in that chapter, along with a simple definition for each word or phrase. As you read the text, you will find these words printed in **boldface type** the first time they appear. You may want to refer to the Key Words list at the beginning of the chapter when you come to a **boldface** word, to learn the definition. At the end of the chapter you may also want to go back and review those definitions. Appendix C is a glossary of all the key words used in the book. They are arranged in alphabetical order for your convenience as you review before the exam.

As you study the material you will be instructed to turn to sections of the question pool in Chapter 10. Be sure to use these questions to review your understanding of the material at the suggested times. This will break the material into bite-sized pieces, and make it easier for you to learn. Do not try to memorize all 326 questions. That will be impossible! Instead, by using the questions for review, you will be familiar with them when you take the test, but you will also understand the electronics theory behind the questions.

Most people learn more when they are actively involved in the learning process. Turning to the questions and answers helps you be actively involved, and will help you learn faster. If you would like the correct answers marked in the question pool, you can mark them as you study. This will reinforce the material in your mind. Make an asterisk or check mark in the left margin next to the correct answer. Then you can cover those marks with a slip of paper for review later. Paper clips make excellent place markers to help you find your spot in the text, question pool and answer key.

The correct answers to the questions are given in the answer key printed at the end of Chapter 10. Also printed with the answer key are page references indicating where you can turn to the text for a quick review of the material that explains each question.

In addition to this book, you will want to purchase a copy of *The FCC Rule Book*, published by the ARRL, which covers all the rules and regulations you'll need to know.

If you have not passed the 5 WPM Morse code exam, you will have to learn the code and increase your speed to 5 WPM. ARRL offers a complete set of cassette tapes to help you with the code. Even with the tapes, you'll want to tune in to the code-practice sessions transmitted by W1AW, the ARRL Headquarters station. The W1AW code-practice sessions are transmitted to assist you in learning the code. For more information about W1AW or how to order any ARRL publication or code tape, write to: ARRL Headquarters, 225 Main St, Newington, CT 06111.

Table of Contents

Chapter 1

The Technician Class License

Many people regard the Technician license as a good starting point in Amateur Radio. After all, with a Technician license comes the privilege of operating through repeaters on the popular 2-meter band.

If you have put off earning your Technician license because you thought the electronics theory was too difficult, put those thoughts aside right now! Once you make the commitment to study and learn what it takes to pass the exam, you *will* be able to do it. It often takes more than one try to pass the exam, but many amateurs do pass on their first try. The key is that you must make the commitment, and be willing to study. To build your code speed there are many good Morse code training techniques, including the ARRL code tapes, W1AW code practice and even some computer programs. With this book, carefully designed to teach the required electronics theory, *The FCC Rule Book* published by the ARRL, and plenty of code-practice, you will soon have a higher-class license. You will find the operating privileges available to a Technician class licensee to be worth the time spent learning about your hobby.

With a Novice or Technician class license, you can "work the world" on segments of four different high-frequency (HF) bands, using the ham's special language, international Morse code. Novice and Technician licensees may also use single-sideband voice and digital modes on the 10-meter band. In addition, Technicians may operate on all frequency bands allocated to the Amateur Radio Service above the HF bands (and there are a lot of them!), ranging from the 6-meter and popular 2-meter bands to the upper reaches of the frequency spectrum, where amateur work is largely experimental.

You can see that the Technician class license allows a wide range of transmitting privileges. To earn those privileges, however, you will have to demonstrate that you know the international Morse code, basic electronic theory, operating practices, and FCC rules and regulations. This book is designed to teach you what you need to know to qualify for the Technician class Amateur Radio license.

IF YOU'RE A NEWCOMER TO AMATEUR RADIO

Earning an Amateur Radio license, at whatever level, is a special achievement. The half a million or so people in the US who call themselves Amateur Radio operators, or hams, are part of a global fraternity. Hams have fun operating two-way radio stations from their homes, cars and the out-of-doors. They talk with other hams across town and around the world using even the simplest of radio setups and antennas.

Just about anyone can be a ham; women, men, boys and girls. Hams range in age from under 8 to over 100! There are no limits. Hams come from all walks of life. Many famous people are hams, but the majority are ordinary folks who like making new friends every day. Nearly anyone can use Amateur Radio to open their door to the world.

Amateur Radio is primarily a hobby for personal enjoyment. But, in times of need, radio amateurs serve the public as a voluntary, noncommercial, communication service. This is especially true during natural disasters or other emergencies.

In addition, hams continue to make important contributions to the field of electronics. Amateur Radio experimentation is yet another reason many people become part of this self-disciplined group of trained operators, technicians and electronics experts—an asset to any country. Hams pursue their hobby purely for personal enrichment in technical and operating skills, without consideration of any type of payment.

Because radio signals do not know territorial boundaries, hams have a unique ability (and responsibility) to enhance international goodwill. A ham becomes an ambassador of his country every time he puts his station on the air.

Amateur Radio has been around since before World War I. Hams have always been at the forefront of technology. Today hams relay signals through their own satellites in the OSCAR (Orbiting Satellite Carrying Amateur Radio) series. They bounce signals off the moon and use any number of other communications techniques. Amateurs talk from hand-held transceivers through mountaintop repeater stations. These repeaters relay their signals to transceivers in other hams' cars or homes. Hams send pictures by television, chat with other hams around the world by voice or tap out messages in Morse code. When emergencies arise, radio amateurs are on the spot to relay information to and from disaster-stricken areas that have lost normal lines of communication.

The US government, through the Federal Communications Commission (FCC), grants all US Amateur Radio licenses. This licensing procedure ensures that radio amateurs have the necessary operating skill and electronics knowledge. Without these skills, radio operators might unknowingly cause interference to other services using the radio spectrum because of improperly adjusted equipment or neglected regulations.

Who Can Be a Ham?

The FCC doesn't care how old you are or whether you're a US citizen: If you pass the examination, the Commission will issue you an amateur license. Any person (except the agent of a foreign government) may take an exam, and, if successful, receive an amateur license. It's important to understand that if a citizen of a foreign country receives an amateur license in this manner, he or she is a US Amateur Radio operator. (You should not confuse this with reciprocal licensing. This agreement allows visitors from certain countries who hold valid amateur licenses in their homelands to operate their own stations in the US without having to take an FCC exam.)

Licensing Structure

By examining Table 1-1, you'll see that there are five US amateur license classes. Each class has its own requirements and privileges. The FCC requires proof of your ability to operate an amateur station properly. The required knowledge is in line with the privileges of the license you hold. Higher license classes require more knowledge—and offer greater operating privileges. The specific operating privileges for Technician class licensees are listed in Table 1-2. As you upgrade your license class, you must pass more challenging written examinations.

In addition, you must show an ability to receive international Morse code at 5 words per minute (WPM) for Novice and Technician, 13 WPM for General and Advanced,

Table 1-1

Amateur Operator Licenses†

Class	Code Test	Written Examination	Privileges
Novice	5 WPM (Element 1A)	Elementary theory and regulations (Element 2)	Telegraphy on 3700-3750, 7100-7150 and 21,100-21,200 kHz with 200 watts PEP output maximum; telegraphy and RTTY on 28,100-28,300 kHz and telegraphy and SSB voice on 28,300-28,500 kHz with 200 W PEP max; all amateur modes authorized on 222.1-223.91 MHz, 25 W PEP max; all amateur modes authorized on 1270-1295 MHz, 5 W PEP max.
Technician	5 WPM (Element 1A)	Elementary theory and regulations; technician-level theory and regulations. Elements 2 and 3A)	All amateur privileges above 50.0 MHz plus Novice HF privileges.
General	13 WPM (Element 1B)	Elementary theory and regulations; technician and general theory and regulations. (Elements 2, 3A and 3B)	All amateur privileges except those reserved for Advanced and Amateur Extra class; see Table 1-2
Advanced	13 WPM (Element 1B)	All lower exam elements, plus intermediate theory. (Elements 2, 3A, 3B and 4A)	All amateur privileges except those reserved for Advanced and Amateur Extra class; see Table 1-2.
Amateur Extra	20 WPM (Element 1C)	All lower exam elements plus special exam on advanced techniques (Elements 2, 3A, 3B, 4A and 4B	All amateur privileges

†A licensed radio amateur will be required to pass only those elements that are not included in the examination for the amateur license currently held.

and 20 WPM for Amateur Extra. (Every five letters count as a word. Numbers, punctuation and procedural signals count as two letters.) Although you may intend to use voice rather than code, it is important to stress that this doesn't excuse you from the code test. By international treaty, knowing the international Morse code is a basic requirement for operating on any amateur band below 30 MHz.

Learning the Morse code is a matter of practice. Instructions on learning the code, how to handle a telegraph key and so on can be found in the ARRL's *Tune in the World with Ham Radio* package. This package includes two code-teaching cassettes for beginners. (Those cassettes are also available separately from *Tune in the World.*) Additional cassettes for code practice at speeds between 5 and 10 WPM, 10 and 15 WPM, and 15 and 22 WPM are available from the American Radio Relay League,

Table 1-2
Amateur Operating Privileges

160 METERS

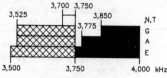

1,800   1,900   2,000 kHz

Amateur stations operating at 1900–2000 kHz must not cause interference to the radiolocation service and are afforded no protection from radiolocation operations.

KEY

⬚ (cross-hatch) = CW and RTTY
⬚ (light shade) = CW, VOICE, SSTV, FAX, and RTTY
■ (black) = CW, VOICE, SSTV, and FAX
▨ (diagonal) = CW and SSB
☐ = CW only

E = EXTRA
A = ADVANCED
G = GENERAL
T = TECHNICIAN
N = NOVICE

80 METERS

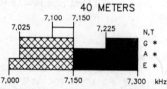

3,525   3,700   3,750   3,775   3,850

N,T / G / A / E

3,500   3,750   4,000 kHz

5,167.5 kHz (SSB only) Alaska emergency use only.

40 METERS

7,025   7,100   7,150   7,225

N,T / G * / A * / E *

7,000   7,150   7,300 kHz

* Phone operation is allowed on 7075–7100 kHz in Puerto Rico, US Virgin Islands and areas of the Caribbean south of 20 degrees north latitude; and in Hawaii and areas near ITU Region 3, including Alaska.

30 METERS

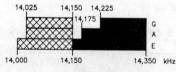

E,A,G

10,100   10,150 kHz

Maximum power on 30 meters is 200 watts PEP output. Amateurs must avoid interference to the fixed service outside the US.

20 METERS

14,025   14,150   14,175   14,225

G / A / E

14,000   14,150   14,350 kHz

17 METERS

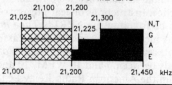

E,A,G

18,068   18,110   18,168 kHz

15 METERS

21,025   21,100   21,200   21,225   21,300

N,T / G / A / E

21,000   21,200   21,450 kHz

12 METERS

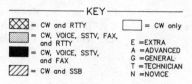

E,A,G

24,890   24,930   24,990 kHz

10 METERS

28,100   28,500

N,T
E,A,G

28,000   28,300   29,700 kHz

Novices and Technicians are limited to 200 watts PEP output on 10 meters.

6 METERS

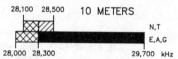

50.1

E,A,G,T

50.0   54.0 MHz

2 METERS

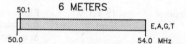

144.1

E,A,G,T

144.0   148.0 MHz

1.25 METERS

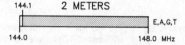

222.1   223.91

N
E,A,G,T

220.0   225.0 MHz

Novices are limited to 25 watts PEP output from 222.1 to 223.91 MHz.

70 CENTIMETERS

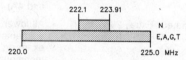

E,A,G,T

420.0   450.0 MHz

33 CENTIMETERS

E,A,G,T

902.0   928.0 MHz

23 CENTIMETERS

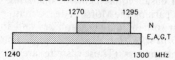

1270   1295

N
E,A,G,T

1240   1300 MHz

Novices are limited to 5 watts PEP output from 1270 to 1295 MHz.

Newington, CT 06111. These tapes use a character speed of 18 WPM with additional space between letters and words to slow the overall speed for practice at less than 18 WPM. In addition there is a set of code practice tapes at 13 to 14 WPM using standard code timing as preparation for the 13 WPM General class code exam.

Code Practice

Besides listening to code tapes, on-the-air operating experience is a great help in building your code speed. When you have to copy the code the other station is sending to continue the conversation, your copying ability will improve quickly!

ARRL's Hiram Percy Maxim Memorial Station, W1AW, transmits code practice and bulletins of information of interest to all amateurs. These code-practice sessions and Morse code bulletins provide an excellent opportunity for code practice. Table 1-3 is an abbreviated W1AW operating schedule. When we change from Standard Time to Daylight Saving Time, the same local times are used.

Station Call Signs

Many years ago most of the world's nations agreed to allocate certain call-sign prefixes to each country. This means that if you hear a radio station call sign beginning with W or K, for example, you know the station is licensed by the United States. A call sign beginning with the letter U is licensed by the USSR, and so on.

International Telecommunication Union (ITU) radio regulations outline the basic principles used in forming amateur call signs. According to these regulations, an amateur call sign must be made of one or two characters (either of which may be a numeral) as a prefix. This is followed by a numeral and then a suffix of not more than three letters. The prefixes W, K, N and A are used in the United States. The continental US is divided into 10 Amateur Radio call districts (sometimes called areas), numbered 0 through 9. Figure 1-1 is a map showing the US call districts.

All US Amateur Radio call signs assigned by the FCC after March 1978 can be categorized into one of five groups, each corresponding to a class, or classes, of license. Call signs are issued systematically by the FCC; requests for special call signs are not granted. For further information on the FCC's call-sign assignment system, and a table listing the blocks of call signs for each license class, see the ARRL publication, *The FCC Rule Book*.

If you already have an amateur call sign, you may keep the same one when you change license class if you wish. You must indicate your preference to receive a new call sign when you fill out an FCC Form 610 to apply for the exam or change your address.

EARNING A LICENSE

Applying for an Exam: FCC Form 610

There are three different types of FCC 610 forms that you may need to use. For most purposes, an ordinary FCC Form 610 will suit your needs. Alien applications for reciprocal operating permission in the US must be made on FCC Form 610-A. Club, military recreation and RACES station license renewal or modification applications must be made on FCC Form 610-B.

Amateurs and other persons must use Form 610 to renew their primary station and operator licenses, make modifications such as address or call sign changes, apply for reinstatement of licenses that have expired within the past two years and to request new Novice licenses. Be sure to use a valid edition of the Form 610. Using an out-of-date form will delay issuance of your license. Your application will be returned without action and you will be required to refile on a current form.

Form 610's showing a revision date earlier than June, 1984 are no longer valid,

Table 1-3

W1AW Schedule

MTWThFSSn = Days of Week   Dy = Daily

W1AW code practice and bulletin transmissions are sent on the following schedule:

EST	Slow Code Practice	MWF: 9AM, 7PM; TThSSn: 4 PM, 10 PM
	Fast Code Practice	MWF: 4 PM, 10 PM; TTh: 9AM; TThSSn: 7PM
	CW Bulletins	Dy: 5 PM, 8 PM, 11 PM; MTWThF: 10 AM
	Teleprinter Bulletins	Dy: 6PM, 9 PM, 12 PM; MTWThF: 11 AM
	Voice Bulletins	Dy: 9:30 PM, 12:30 AM
CST	Slow Code Practice	MWF: 8 AM, 6 PM; TThSSn: 3 PM, 9 PM
	Fast Code Practice	MWF: 3 PM, 9 PM; TTh: 8 AM; TThSSn: 6 PM
	CW Bulletins	Dy: 4 PM, 7 PM, 10 PM; MTWThF: 9 AM
	Teleprinter Bulletins	DY: 5 PM, 8 PM, 11 PM; MTWThF: 10 AM
	Voice Bulletins	Dy: 8:30 PM, 11:30 PM
MST	Slow Code Practice	MWF: 7 AM, 5 PM; TThSSn: 2 PM; 8 PM
	Fast Code Practice	MWF: 2 PM, 8 PM; TTh: 7 AM; TThSSn: 5 PM
	CW Bulletins	Dy: 3 PM, 6 PM, 9 PM; MTWThF: 8 AM
	Teleprinter Bulletins	Dy: 4 PM, 7 PM, 10 PM; MTWThF: 9 AM
	Voice Bulletins	Dy: 7:30 PM, 10:30 PM
PST	Slow Code Practice	MWF: 6 AM, 4 PM; TThSSn: 1 PM; 7 PM
	Fast Code Practice	MWF: 1 PM, 7 PM; TTh: 6 AM; TThSSn: 4 PM
	CW Bulletins	Dy: 2 PM, 5 PM, 8 PM; MTWThF: 7 AM
	Teleprinter Bulletins	Dy: 3 PM, 6 PM, 9 PM; MTWThF: 8 AM
	Voice Bulletins	Dy: 6:30 PM, 9:30 PM

Code practice, Qualifying Run and CW bulletin frequencies: I.8l8, 3.58, 7.08, 14.07, 2l.08, 28.08, 50.08, l47.555 MHz.

Teleprinter bulletin frequencies: 3.625, 7.095, 14.095, 21.095, 28.095, 147.555 MHz. Voice bulletin frequencies: I.89, 3.99, 7.29, 14.29, 2l.39, 28.59, 50.19, 147.555 MHz.

On Monday, Wednesday and Friday, 9 AM through 5 PM EST, transmissions are beamed to Europe on 14, 21 and 28 MHz; on Wednesday at 6 PM EST they are beamed south.

Slow code practice is at 5, 7½, 10, 13 and 15 WPM. Fast code practice is at 35, 30, 25, 20, 15, 13 and 10 WPM.

Code practice texts are from *QST*, and the source of each practice is given at the beginning of each practice and at the beginning of alternate speeds. For example, "Text is from July 1985 *QST*, pages 9 and 76" indicates that the main text is from the article on page 9 and the mixed number/letter groups at the end of each speed are from the contest scores on page 76.

Some of the slow practice sessions are sent with each line of text from *QST* reversed. For example, "Last October, the ARRL Board of Directors" would be sent as DIRECTORS OF BOARD ARRL THE, OCTOBER LAST.

W1AW CW and voice bulletins are sent on OSCAR 10, Mode B, when the satellite is within range. Look for CW on 145.840 MHz and SSB on 145.962 MHz.

Teleprinter bulletins are 45.45-baud Baudot, 110-baud ASCII and 100-baud AMTOR, FEC mode. Baudot, ASCII and AMTOR (in that order) are sent during all 11 AM EST transmissions, and 6 PM EST on TThFSSn. During other transmission times, AMTOR is sent only as time permits.

CW bulletins are sent at 18 WPM.

W1AW is open for visitors Monday through Friday from 8 AM to 1 AM EST and on Saturday and Sunday from 3:30 PM to 1 AM EST. If you desire to operate W1AW, be sure to bring a copy of your license with you. W1AW is available for operation by visitors between 1 and 4 PM Monday through Friday.

In a communications emergency, monitor W1AW for special bulletins as follows: voice on the hour, teleprinter at 15 minutes past the hour, and CW on the half hour.

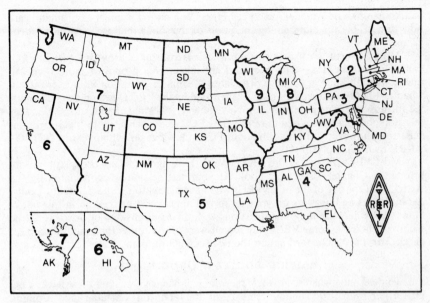

Figure 1-1—There are 10 US call areas. Hawaii is part of the sixth call area, and Alaska is part of the seventh.

and will be returned if you attempt to use one. If you have any old forms, throw them away. (At the time of this writing, there are five versions of the FCC Form 610 that are valid. Only four show release dates in the lower right corner of the form: June 1984, July 1985, March 1986 and September 1987. The other version, virtually identical to the September 1987 form, was released in February 1987, but it does not show a release date. This form has an expiration date of 12/31/89 printed in the upper right corner, however, as does the September 1987 one.) All but the latest two versions will require some modifications to the form when you fill it out. You would do well to obtain a copy of the 610 form that has an expiration date of 12/31/89 or later.

To obtain a new Form 610, 610-A or 610-B, send a business-sized, self-addressed, stamped envelope to: Federal Communications Commission, Form 610, PO Box 1020, Gettysburg, PA 17326 or: Form 610, ARRL, 225 Main St., Newington, CT 06111.

Volunteer-Examiner Program

Since January 1, 1985, all US amateur exams above the Novice level have been administered under the Volunteer-Examiner Program. *The FCC Rule Book* contains details on this program. Novice exams have always been given by volunteer examiners, and that is still true. Those exams do not come under the regulations involving Volunteer-Examiner Coordinators (VECs), however.

To qualify for a Technician class license, you must pass FCC Exam Element 1A, Element 2 and Element 3A. If you already hold a valid Novice license, then you have credit for passing Elements 1A and 2, and will not have to retake them when you go for your Technician class exam. See Table 1-1 for details.

The FCC requires a minimum of 25 questions for an Element 3A exam, along with a percentage of questions from each subelement that must appear on the exam. The FCC specifies nine subelements, or divisions, for each exam element. These are the Commission's Rules, Operating Procedures, Radio-Wave Propagation, Amateur Radio Practices, Electrical Principles, Circuit Components, Practical Circuits, Signals and Emissions, and Antennas and Feed Lines. Volunteer-Examiner Coordinators are now responsible for maintaining the pools of questions that are used on amateur exams.

Most Element 3A exams (including exams coordinated by the ARRL/VEC) consist of 25 questions. The questions are taken from a pool of more than 250 questions, published in advance. The latest Technician Question Pool is printed in Chapter 10 of this book. This pool was released by the VEC Question Pool Committee for use on exams given between November 1, 1989 and October 31, 1992. This question pool may be revised as required by FCC rules changes or to fix typographical errors, however. Information about FCC rules changes will appear in *QST*, the ARRL membership journal.

Most VECs, including the ARRL/VEC, have agreed to use the question pool printed in Chapter 10 of this book. If your test session is coordinated by the ARRL/VEC or one of the other VECs who have agreed to use these question pools, the questions and answers on your exam will appear just as they do in this book. Some VECs may use the same questions with different answers or distractors (the incorrect choices). Some VECs may use answer formats other than multiple choice; check with the VEC coordinating the test session you plan to attend.

Finding an Exam Opportunity

To determine where and when an exam will be given, contact the ARRL/VEC Office, or watch for announcements in the Hamfest Calendar and Coming Conventions columns in *QST*. Many local clubs sponsor exams, so they are another good source of information on exam opportunities. ARRL officials such as Directors, Vice Directors and Section Managers receive notices about test sessions in their area. See page 8 in the latest issue of *QST* for names and addresses.

To register for an exam, send a completed Form 610 to the volunteer-Examiner Team responsible for the exam session if preregistration is required. Otherwise, bring the form to the session. Registration deadlines, and the time and location of the exams, are mentioned prominently in publicity releases about upcoming sessions.

Taking The Exam

By the time examination day rolls around, you should have already prepared yourself. This means getting your schedule, supplies and mental attitude ready. Plan your schedule so you'll get to the examination site with plenty of time to spare. There's no harm in being early. In fact, you might have time to discuss hamming with another applicant, which is a great way to calm pretest nerves. Try not to discuss the material that will be on the examination as this may make you even more nervous. By this time, it's too late to study anyway!

What supplies will you need? First, be sure you bring your current *original* Amateur Radio license, if you have one. Bring along several sharpened number 2 pencils and two pens (blue or black ink). Be sure to have a good eraser. A pocket calculator may also come in handy. You may use a programmable calculator if that is the kind you have, but take it into your exam "empty" (cleared of all programs and constants in memory). Don't program equations ahead of time because you may be asked to demonstrate that there is nothing in the calculator's memory. If you use a slide rule, that should also be allowed. You will probably *not* be allowed to take math tables, such as trigonometery tables or logarithim tables into the exam with you.

The Volunteer-Examiner Team is required to check two forms of identification before you enter the test room. This includes your *original* Amateur Radio license. A photo ID of some type is best for the second form, but is not required by FCC. Other acceptable forms of identification include a driver's license, a piece of mail addressed to you, a birth certificate or some other such document.

The following description of the testing procedure applies to exams coordinated by the ARRL/VEC, although many other VECs use a similar procedure. The code tests are usually given before the written exams. The 20 WPM exam is usually given first, then the 13 WPM exam and finally the 5 WPM test. There is no harm in trying the 20 or 13 WPM exams even though you are testing for the Technician exams.

Before you take the code test, you'll be handed a piece of paper to copy the code as it's sent. The test will begin with about a minute of practice copy. Then comes the actual test: five minutes of Morse code. You are responsible for knowing the 26 letters of the alphabet, the numerals 0 through 9, the period, comma, question mark, $\overline{AR}$, $\overline{SK}$, $\overline{BT}$ and $\overline{DN}$. You may copy the entire text word for word, or just take notes on the content. At the end of the transmission, the examiner will hand you 10 questions about the text. Simply fill in the blanks with your answers. (You must spell each answer exactly as it was sent.) If you get at least 7 correct, you pass! Alternatively, the exam team has the option to look at your copy sheet if you fail the 10-question exam. If you have one minute of solid copy, they can certify that you passed the test on that basis. The format of the test transmission is similar to one side of a normal on-the-air amateur conversation.

A sending test may not be required. The Commission has decided that if applicants can demonstrate receiving ability, they most likely can also send at that speed. But be prepared for a sending test, just in case! Subpart 97.21(a) of the FCC rules says, "A telegraphy examination shall be such as to prove that a person has the ability to send correctly by hand and receive correctly by ear texts in the international Morse code at the speed listed for the appropriate examination element."

If all has gone well with the code test, you'll then take the written examination. The examiner will give each applicant a test booklet, an answer sheet and scratch paper. After that, you're on your own. The first thing to do is read the instructions. Be sure to sign your name every place it's called for. Do all of this at the beginning to get it out of the way.

Next, check the examination to see that all pages and questions are there. If not, report this to the examiner immediately. When filling in your answer sheet, make sure your answers are marked next to the numbers that correspond to each question.

Go through the entire exam, and answer the easy questions first. Next, go back to the beginning and try the harder questions. The really tough questions should be left for last. Guessing can only help, as there is no additional penalty for answering incorrectly.

If you have to guess, do it intelligently: At first glance, you may find that you can eliminate one or more "distractors." Of the remaining responses, more than one may seem correct; only one is the best answer, however. To the applicant who is fully prepared, incorrect distractors to each question are obvious. Nothing beats preparation!

After you've finished, check the examination thoroughly. You may have read a question wrong or goofed in your arithmetic. Don't be overconfident. There's no rush, so take your time. Think, and check your answer sheet. When you feel you've done your best and can do no more, return the test booklet, answer sheet and scratch pad to the examiner.

The Volunteer-Examiner Team will grade the exam right away. The passing mark is 74%. (That means no more than 6 incorrect answers on a 25 question exam.) If you are already licensed, and you pass the exam elements required to earn a higher class of license, you will receive a Certificate of Successful Completion of Examination (CSCE) which allows you to operate with your new privileges. The certificate has a special identifier code that must be used on the air when you use your new privileges, until your permanent license arrives from the FCC.

If you pass only some of the exam elements required for a higher class license you will still receive a CSCE. That certificate shows what exam elements you passed, and is valid for one year. Use it as proof that you passed those exam elements so you won't have to take them over again next time you try for the upgrade.

AND NOW, LET'S BEGIN

Subelement 3AA in Chapter 10 covers the rules and regulations for the Element 3A exam. You should use *The FCC Rule Book* to find the material covered by those questions. Then go over that section of the question pool to check your understanding of the rules. Perhaps you will want to study the rules a few at a time, using that as a break from your study in the rest of this book.

There you have it. The remainder of this book will provide the background in electronics theory that you will need to pass the Element 3A Technician class written exam. Table 1-4 shows the study guide or syllabus for the Element 3A exam. This study guide was released by the Volunteer Examiner Coordinators' Question Pool Committee in July 1988. The syllabus lists the topics to be covered by the Technician exam, and so forms a basic outline for the remainder of this book. Use the syllabus to guide your study, and to ensure that you have studied the material for all of the topics listed.

This syllabus is organized as an outline, and the points are numbered in outline form. The question numbers used in the question pool refer to this syllabus. Each question number begins with a syllabus-point number (the part to the left of the decimal point) and ends with a serial number that identifies a specific question. For example, question 3BA-2-1.5 is the fifth question related to syllabus point 3BA-2-1, use of a repeater.

Table 1-4
Element 3A (Technician) Syllabus
(Required for all operator licenses above Novice)

Subelement 3AA Commission's Rules
(5 Exam Questions)
3AA-1 Control point, definition {97.3 (p)}
3AA-2 Frequency privileges for Technician class control operators {97.7(b), (f)}
3AA-3 Renewal or modification of operator and station licenses {97.13; 97.47}
3AA-4 Emission privileges for Technician class control operators {97.61}
3AA-5 Selection and use of frequencies {97.63}
3AA-6 Transmitter power
   3AA-6-1 Definition {97.3(t)(1)}
   3AA-6-2 Minimum power necessary {97.67(a)}
   3AA-6-3 Maximum power permitted {97.67(b), (d)}
   3AA-6-4 Power of station in beacon operation {97.67(e)}
3AA-7 Digital communications
   3AA-7-1 Maximum sending speed {97.69(a)(1)}
   3AA-7-2 Maximum frequency shift {97.69(a)(2)}
   3AA-7-3 Maximum permitted bandwidth above 50 MHz {97.69(c)(2)}
3AA-8 Station identification

3AA-8-1 Operating with a Certificate of Successful Completion of Examination {97.35; 97.84(f)}
   3AA-8-2 Telephony (what language to use) {97.84(g)(2)}
   3AA-8-3 Phonetic alphabet {97.84(g)(2)}
3AA-9 Beacon operation
   3AA-9-1 Definition {97.3(l)}
   3AA-9-2 Class of license required {97.87(f)}
3AA-10 Radio control of model craft and vehicles {97.3(l); 97.99}
3AA-11 Emergency communications
   3AA-11-1 Definition {97.3(w)}
   3AA-11-2 Declaration of general state of communications emergency {97.107}
3AA-12 Broadcasting {97.113(a), (b), (c)}
3AA-13 Permissible one-way transmissions {97.113(d), (e)}
3AA-14 Domestic and international third-party traffic {97.114(a), (b)}
3AA-15 Third-party participation {97.114(c)}
3AA-16 Transmission of indecent or profane language {97.119}
3AA-17 Communication through satellites {97.409}

Subelement 3AB Operating Procedures
(3 Exam Questions)

3AB-1 RST signal reporting system
3AB-2 Repeater operation
   3AB-2-1 Use of a repeater
   3AB-2-2 Repeater versus simplex operation
   3AB-2-3 Input/output frequency separation (Frequency splits used on various repeater subbands)
   3AB-2-4 Frequency coordination
3AB-3 Operating courtesy
3AB-4 Distress calling procedures
3AB-5 Emergency preparedness drills
   3AB-5-1 RACES drills
   3AB-5-2 Messages identified as "drill" or "test"
3AB-6 Providing emergency communications
   3AB-6-1 Tactical communications
   3AB-6-2 Health and welfare traffic
   3AB-6-3 Equipment considerations

Subelement 3AC Radio-Wave Propagation
(3 Exam Questions)

3AC-1 Ionosphere
   3AC-1-1 Definition
   3AC-1-2 D layer
   3AC-1-3 E layer
   3AC-1-4 F layers
3AC-2 Ionospheric absorption
3AC-3 Daily variation in ionization levels of ionosphere
3AC-4 Maximum usable frequency
3AC-5 Scatter propagation
3AC-6 Line-of-sight propagation
3AC-7 Tropospheric bending and ducting

Subelement 3AD Amateur Radio Practices
(4 Exam Questions)

3AD-1 Electrical wiring safety
  3AD-1-1 Wiring polarity
  3AD-1-2 Dangerous voltages and currents
  3AD-1-3 Placement and rating of fuses and switches
3AD-2 Voltmeters
  3AD-2-1 Connection in circuit
  3AD-2-2 Extending range of meters
3AD-3 Ammeters
  3AD-3-1 Connection in circuit
  3AD-3-2 Extending range of meters
3AD-4 Multimeters
3AD-5 Wattmeters
  3AD-5-1 Connection in circuit
  3AD-5-2 Interpreting measurements
3AD-6 Marker generators
3AD-7 Signal generators
3AD-8 Impedance-match indicators
  3AD-8-1 SWR meters
  3AD-8-2 Placement in feed line
3AD-9 Dummy antennas
3AD-10 Use of S meters
3AD-11 RF Safety
  3AD-11-1 Thermal effects of RF on the body
  3AD-11-2 American National Standards Institute (ANSI) RF protection guide
    (Limited to the concept that a guide exists and that it sets exposure limits
    under certain circumstances)
  3AD-11-3 Minimizing exposure to RF

Subelement 3AE Electrical Principles
(2 Exam Questions)

3AE-1 Resistance
  3AE-1-1 Definition
  3AE-1-2 Units
  3AE-1-3 Resistors in series
  3AE-1-4 Resistors in parallel
3AE-2 Ohm's Law
3AE-3 Inductance
  3AE-3-1 Definition
  3AE-3-2 Units
  3AE-3-3 Inductors in series
  3AE-3-4 Inductors in parallel
3AE-4 Capacitance
  3AE-4-1 Definition
  3AE-4-2 Units
  3AE-4-3 Capacitors in series
  3AE-4-4 Capacitors in parallel

Subelement 3AF Circuit Components
(2 Exam Questions)

3AF-1 Resistors
    3AF-1-1 Construction types
    3AF-1-2 Variable and fixed resistors
    3AF-1-3 Color code
    3AF-1-4 Power rating
    3AF-1-5 Schematic symbols
3AF-2 Inductors
    3AF-2-1 Construction
    3AF-2-2 Variable and fixed inductors
    3AF-2-3 Factors affecting inductance
    3AF-2-4 Schematic symbols
3AF-3 Capacitors
    3AF-3-1 Construction
    3AF-3-2 Variable and fixed capacitors
    3AF-3-3 Factors affecting capacitance
    3AF-3-4 Schematic symbols

Subelement 3AG Practical Circuits
(1 Exam Question)

3AG-1 Low-pass filters
    3AG-1-1 Frequency characteristics
    3AG-1-2 Applications
3AG-2 High-pass filters
    3AG-2-1 Frequency characteristics
    3AG-2-2 Applications
3AG-3 Band-pass filters
    3AG-3-1 Frequency characteristics
    3AG-3-2 Applications
3AG-4 Transmitter and receiver block diagrams (Know functions of various blocks and how they work together)
    3AG-4-1 A1A transmitters and receivers
    3AG-4-2 F3E transmitters and receivers

Subelement 3AH Signals and Emissions
(2 Exam Questions)

3AH-1 Modulation (definition)
3AH-2 Emission types
    3AH-2-1 NØN
    3AH-2-2 A1A
    3AH-2-3 F1B
    3AH-2-4 F2A
    3AH-2-5 F2B
    3AH-2-6 F2D
    3AH-2-7 F3E
    3AH-2-8 G3E

3AH-3 RF carrier
3AH-4 Frequency modulation
3AH-5 Phase modulation
3AH-6 Bandwidth (Limited to concept that different emissions occupy different bandwidths)
3AH-7 Deviation
   3AH-7-1 Relation to audio modulating signal
   3AH-7-2 Overdeviation

Subelement 3AI Antennas and Feed Lines (3 Exam Questions)

3AI-1 Parasitic beam antennas
   3AI-1-1 Yagi antennas
   3AI-1-2 Quad antennas
   3AI-1-3 Delta loop antennas
3AI-2 Polarization of antennas and radio waves
   3AI-2-1 Horizontal polarization
   3AI-2-2 Vertical polarization
3AI-3 Standing wave ratio (SWR)
   3AI-3-1 Definition
   3AI-3-2 Forward and reflected power
   3AI-3-3 Significance of SWR to system
3AI-4 Balanced and unbalanced conditions
   3AI-4-1 Feed lines
   3AI-4-2 Antennas
   3AI-4-3 Balun transformers
3AI-5 Feed-line attenuation
   3AI-5-1 Changes with line type
   3AI-5-2 Changes with line length
   3AI-5-3 Changes with frequency
3AI-6 RF safety
   3AI-6-1 Feed lines
   3AI-6-2 Antennas

Key Words

Band plan —An agreement for operating within a certain portion of the radio spectrum. Band plans set aside certain frequencies for each different mode of amateur operation, such as CW, SSB, FM, repeaters and simplex.

Clipping—Occurs when the peaks of a voice waveform are cut off in a transmitter, usually because of overmodulation. Also called **flattopping**.

Frequency Coordinator—A volunteer who keeps records of repeater input, output and control frequencies.

MAYDAY—From the French "m'aider" (help me), MAYDAY is used when calling for emergency assistance in voice modes.

Monitor Oscilloscope—A test instrument connected to an amateur transmitter and used to observe the shape of the transmitted-signal waveform.

Offset—The difference between a repeater's input and output frequencies. On 2 meters, for example, the offset is either plus 600 kilohertz (kHz) or minus 600 kHz from the receive frequency.

RST System—The system used by amateurs for giving signal reports. "R" stands for readability, "S" for strength and "T" for tone. See Table 2-1.

Simplex—A term normally used in relation to VHF and UHF operation, simplex operation means you are receiving and transmitting on the same frequency.

SOS—A Morse code call for emergency assistance.

Splatter—The term used to describe a very wide bandwidth signal, usually caused by overmodulation of a sideband transmitter. Splatter causes interference to adjacent signals.

Chapter 2

Operating Procedures

T he thrill of Amateur Radio is on-the-air operation. There are many different operating modes available. True enjoyment of the hobby comes from using this diversity. This chapter covers on-the-air procedures that you will use while operating with Technician class privileges.

TELEPHONY

Radiotelephony is probably the most natural communications mode of all. Phone operation is not as simple as grabbing a microphone and talking, however. Keep in mind a few rules and guidelines for courteous, legal, on-the-air operation. Make sure your transmitter is operating properly. Adjust your microphone gain correctly. This adjustment can be made to an SSB (type J3E emission) transmitter while observing a **monitor oscilloscope**. Proper voice waveforms have rounded peaks. Peaks with flat tops indicate **clipping** (also called **flattopping**). Clipping occurs when voice-waveform peaks are cut off in the transmitter. Excessive microphone gain causes clipping. See Figure 2-1. *The ARRL Handbook* shows the proper oscilloscope connections for monitoring your signal. Consult the owner's manual for your rig about instructions on proper operation and gain-control settings.

If your rig has an automatic level control (ALC) meter, adjust the microphone gain until the ALC meter moves slightly on modulation peaks. With an ALC indicator light, increase the microphone gain until the indicator just lights on voice peaks.

FM transmitters (type F3E emission) should be checked with a deviation meter.

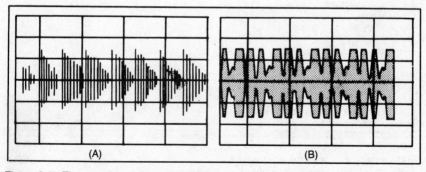

Figure 2-1—The waveform of a properly adjusted SSB transmitter is shown at A. B shows a severely clipped signal.

Use the deviation meter to adjust the transmitter for proper deviation on voice peaks. Once set, FM transmitters usually do not need to be readjusted.

Proper operation is important. Too much microphone gain will distort your transmitted signal. Distortion can cause **splatter**, resulting in interference to stations on other frequencies. If you use a speech processor, use the minimum necessary processing level to limit distortion. An overprocessed, distorted signal is more difficult to copy than an unprocessed one.

Speech is the primary element of radiotelephone. For effective voice communications, your speech should be clearly enunciated. Speak directly into the microphone at a speed suitable for the purpose and conditions at hand. You must use the microphone and other speech equipment properly so other operators can understand you.

Microphones vary in design and purpose. Generally it is best to speak close to the microphone. Hold it a few inches away from your mouth. This will improve the signal-to-noise ratio by making your voice louder than any background noises.

In any mode, (SSB, FM and even AM) pay attention to your mic technique. Keep average speech levels high without overdoing it. Automatic level controls (ALC) help, but you need to maintain relatively constant speech levels.

Signal Reporting

One piece of information that amateurs exchange during almost every QSO is a signal report. A standard system of reporting signal readability, strength and tone is used to describe a CW signal. The **"RST" System** is shown in Table 2-1. R stands

---

Table 2-1
The RST System

Readability:
1-Unreadable
2-Barely readable, occasional words distinguishable
3-Readable with considerable difficulty
4-Readable with practically no difficulty
5-Perfectly readable

Signal Strength:
1-Faint signals barely perceptible
2-Very weak signals
3-Weak signals
4-Readable with practically no difficulty
5-Fairly good signals
6-Good signals
7-Moderately strong signals
8-Strong signals
9-Extremely strong signals

Tone:
1-60-cycle ac or less, very rough and broad
2-Very rough ac, very harsh and broad
3-Rough ac tone, rectified but not filtered
4-Rough note, some trace of filtering
5-Filtered rectified ac but strongly ripple-modulated
6-Filtered tone, definite trace of ripple modulation
7-Near pure tone, trace of ripple modulation
8-Near perfect tone, slight trace of ripple modulation
9-Perfect tone, no trace of ripple or modulation of any kind

---

for readability, S for strength and T for tone. This system has three scales. Use 1 through 5 for readability, 1 through 9 for strength and 1 through 9 for tone. A report of RST 368 would be interpreted as "Your signal is readable with considerable difficulty. It has good strength and a slight trace of modulation." The tone report is a useful indicator of transmitter stability. When the RST system was developed, the tone of amateur transmitters varied widely. Today, a tone report of less than 9 is cause to check your transmitter. Ask a few other amateurs for their opinion of your transmitted signal. Consistent poor tone reports may mean that you have problems.

On phone, the tone report does not apply. Signal reports are given in the Readability-Strength format. A very good signal is 58 or 59. Above S9, strength reports are given as decibels (dB) above S9. A report of 10 dB over S9 means that your signal is 10 dB stronger than an S9 signal.

Don't spend a lot of time worrying about what signal report to give a station. The scales used are rather broad. They are simply a relative indication of how you are receiving the other station. You will be more comfortable at estimating the proper signal report as you gain experience. Use the descriptions shown in Table 2-1.

Give an honest estimate of the other station's signals. Don't simply return the same report you receive. It is entirely possible that other stations are hearing you perfectly, even though you are having trouble hearing them. Your signals could be extremely strong (59), while you have considerable difficulty copying their weak signals (33).

[Now turn to Chapter 10 and study exam questions 3AB-1.1, 3AB-1.2 and 3AB-1.3. Review this section as needed.]

Repeater Operation

A **repeater** is an amateur station that receives a signal and retransmits (repeats) the signal. Repeaters are often located on top of a tall building or a high mountain. They greatly extend the operating coverage of amateurs using mobile and hand-held VHF and UHF transceivers. Most repeaters for voice communication receive on one frequency and simultaneously retransmit on another frequency. Some repeaters (particularly those designed for digital communications) hold the message for transmission later. These repeaters may receive and transmit on the same frequency.

To use a repeater you must have a transceiver that can transmit on the repeater input frequency. This is the frequency the repeater receives on. Your rig must also be able to receive on the repeater output (transmit) frequency. On a modern VHF transceiver using a frequency synthesizer, you simply dial in the correct frequency. Some rigs require setting both the transmit and receive frequencies. On other rigs you set only the receive frequency. Then you move an "offset" switch to the correct position for the transmit frequency. The offset switch controls the separation between receive and transmit frequencies. With older rigs you may have to install new crystals for each desired operating frequency. Consult your owner's manual for complete instructions.

With most repeaters you simply transmit a signal to the repeater, and you will be able to use it. Repeaters like this are known as "open" repeaters, because they are for use by any ham. You will be able to talk to anyone within range of the repeater just by transmitting.

Some repeaters (called "closed" repeaters) require the transmission of a subaudible tone along with the voice signal. Others require a series of tones when you begin to transmit in order to gain access. This type of repeater cannot be accessed simply by keying the microphone. You must transmit the correct subaudible tone or series of tones to get into the repeater.

On most repeaters, the transmitted carrier will continue for a few seconds after you stop transmitting. This is the "squelch tail." If you don't hear a squelch tail at the end of your transmission, you may not have accessed the repeater.

Most repeaters limit the length of a single transmission with a "time-out timer." This ensures that the repeater will turn itself off if the repeater transmitter is on continuously. The timer also prevents one person from monopolizing the repeater. Some repeaters send a short tone as a timer-reset indicator. Others use the Morse code letter K (meaning "go ahead"). Repeaters also use other Morse letters to tell you the timer has reset. Then the next person may transmit without letting the squelch tail drop. On other repeaters you may have to let the squelch tail drop to reset the timer. Ask a regular user of the repeater for proper operating etiquette on a particular machine.

To use repeaters properly, you should learn the operating practices of this unique mode:

1) Monitor the repeater to become familiar with any peculiarities in its operation.

2) To initiate a contact, simply say that you're listening on the frequency. In different geographical areas there are different practices for making yourself heard. In general, something like, "This is KA1IXI monitoring" will suffice. Calling CQ through a repeater is not generally considered good operating practice. If you are looking for a specific person on the repeater, simply call once: "K1XA, this is N2YL."

3) Identify legally. You must transmit your call sign every 10 minutes while you are talking with someone and at the end of the contact.

4) Pause between transmissions. This allows other hams to use the repeater (someone may have an emergency).

5) Keep transmissions short and thoughtful. Your monologue may prevent someone from using the repeater in an emergency.

6) Use **simplex** whenever possible. If you can complete your QSO on a direct frequency, there's no need to tie up the repeater.

7) Use the minimum amount of power necessary to maintain communications. Section 97.67 of the FCC rules requires this. Also, it minimizes the possibility of accessing distant repeaters on the same frequency.

8) Don't break into a contact unless you have something to add. If you wish to join a conversation in progress, wait for a pause in the conversation. Then transmit your call sign once. Use the phrase "break" or "break break" to interrupt a QSO *only* if you have emergency traffic.

9) The FCC forbids using autopatch for anything that could be construed as business communication. An autopatch connects the repeater to the telephone system. For example, don't use the autopatch to call and order a pizza. Nor should you use an autopatch to avoid a toll call. Do not use an autopatch where regular telephone service is available. An autopatch can be very useful in an emergency, however. It can be the fastest way to call police or an ambulance to a highway accident, for example.

10) All repeaters are assembled and maintained at considerable expense and effort. Usually, an individual or group is responsible for the repeater. The regular users of a machine support it. Club dues or other contributions keep the repeater on the air.

The Repeater Directory

The ARRL publishes a *Repeater Directory,* listing over 8000 repeaters in the US and Canada. If you own a rig that operates above 29.5 megahertz (MHz), you'll find *The ARRL Repeater Directory* useful for locating repeaters in your area. It is indispensable when you're traveling.

Frequency coordination

Frequency coordinators are volunteers (not appointed by ARRL) who keep extensive records of repeater input, output and control frequencies. This includes repeaters not listed in directories (sometimes at the owner's request). The coordinator will recommend frequencies for a proposed repeater to minimize interference with other repeaters and simplex frequencies. When a conflict arises, the FCC acknowledges the rights of a coordinated repeater. The commission has taken action against noncoordinated repeaters that caused interference to coordinated repeaters. Anyone thinking about installing a repeater should check with the local frequency coordinator. Check the latest edition of the *Repeater Directory* or contact ARRL Headquarters for the name and address of your area frequency coordinator.

Band Plans

Band plans are agreements among concerned radio amateurs for operating within certain portions of the radio spectrum. The goal is to minimize interference among the various modes sharing each band. This is done by setting aside certain sections of a band for different operating modes. We have discussed FM repeater and simplex activity. There is also CW, SSB, AM, satellite and radio-control operations (among others) on the VHF bands. FCC-mandated subband restrictions apply on the amateur bands between 3.5 and 148 MHz. Different operating modes must share spectrum space. RTTY and CW operators, and slow-scan television and phone operators, for example. Voluntary band plans help minimize conflicts. The ARRL 144- to 148-MHz band plan is shown in Table 2-2. More information on VHF and UHF band plans can be found in *The ARRL Repeater Directory*. HF band plans are in *The ARRL Operating Manual*.

Table 2-2

144-148 MHz

The following band plan has been proposed by the ARRL VHF-UHF Advisory Committee.

144.00-144.05	EME (CW)
144.05-144.06	Propagation beacons
144.06-144.10	General CW and weak signals
144.10-144.20	EME and weak signal SSB
144.20	National calling frequency
144.20-144.30	General SSB operation
144.30-144.50	New OSCAR subband
144.50-144.60	Linear translator inputs
144.50-144.90	FM repeater outputs
145.50-145.80	Miscellaneous and experimental modes
145.80-146.00	OSCAR subband
146.01-146.37	Repeater inputs
146.40-146.58	Simplex
146.61-147.39	Repeater outputs
147.42-147.57	Simplex
147.60-147.99	Repeater inputs

Input/Output Separation

One function of band plans is to specify certain pairs of frequencies for repeater inputs and outputs. The separation between inputs and outputs (called the **offset**) varies from band to band—on 10 meters, the separation is 100 kilohertz (kHz). The K8LK repeater in Doylestown, Ohio uses an input of 29.540 MHz with a 29.640 MHz output. On 6 meters, the split is an even 1 megahertz. The K1FFK repeater on Mount Greylock in Massachusetts has its input on 52.23 MHz and its output on 53.23 MHz.

On 2 meters, the offset is 600 kHz. The W9MQD repeater in Slinger, Wisconsin has its input on 146.13 MHz and its output on 146.73 MHz. Between 144 and 147 MHz, the input of a repeater is usually on the lower frequency and the output on the higher. Between 147 and 148 MHz, the input is usually higher than the output frequency.

On the 220-MHz band (1.25 meters), the separation is 1.6 MHz. You'll find the input of the W1AW 220 machine at 223.24 MHz and the output at 224.84 MHz. Repeaters on the 450-MHz band (70 centimeters) use a 5-MHz split. KØWUG in Willmar, Minnesota has its input on 449.80 MHz and its output on 444.80 MHz.

[Now turn to Chapter 10 and study questions 3AB-2-1.1 through 3AB-2-1.6, questions 3AB-2-3.1 through 3AB-2-3.4 and question 3AB-2-4.1. Review the material in this section as needed.]

GENERAL OPERATING PRACTICES

Operating Courtesy

On today's crowded ham bands, operating courtesy becomes quite important. Operating courtesy is nothing more than common sense. A good operator sets an example for others to follow. A courteous amateur operator always listens before transmitting. This prevents interference with stations already using the frequency. If the frequency is occupied, move to another frequency far enough away to prevent interference (QRM). Know how much bandwidth each mode occupies and act accordingly. A good rule of thumb is to leave 150 to 500 hertz (Hz) between CW (A1A emission) signals. Leave at least 3 kHz between suppressed carriers in single sideband (J3E). Allow 250 to 500 Hz center-to-center for RTTY (F1B) signals.

Good amateur operators transmit only what is necessary to accomplish their purpose. They keep the content of the transmission within the bounds of propriety and good taste. A good operator always reduces power to the minimum necessary to carry out the desired communications. This is not only operating courtesy, it's an FCC rule! Section 97.67 (b) states "...amateur stations shall use the minimum transmitting power necessary to carry out the desired communications." A complete copy of Part 97 appears in *The FCC Rule Book*, published by the ARRL.

The band you select to operate on is important to reduce unintentional QRM. You can cause interference to stations thousands of miles away. Talking to a buddy across town on a band open to distant locations shows inexcusably bad manners. Use VHF and UHF frequencies for local communications.

Sometimes conflicts arise between operators using simplex and a repeater on the VHF bands. It is common courtesy and a simple matter for the simplex operators to change frequency. Changing the operating frequency of a repeater is impractical.

During commuter rush hours, mobile operators should have the highest priority on the repeater. The exception is emergency traffic, of course. All other users, such as third-party traffic nets, should relinquish the repeater to the mobile operators. Repeater operating courtesy includes keeping your transmissions short and to the point. A long-winded monologue may make the repeater shut off (or "time out"). If this does happen the repeater will usually reset itself a short time later. Timing out a repeater can be embarrassing.

Testing and "loading up" a transmitter should be done into a dummy antenna. On-the-air tests should be made only when necessary and kept as brief as possible. When operating fixed or portable within a mile of an FCC monitoring station, contact the staff at the station. This ensures that your amateur operations do not cause them any harmful interference. Always be more courteous than the other person. Listen before transmitting. Ask if the frequency is in use. Send QRL? (Are you busy?) on CW or ask if the frequency is in use on voice. Avoid long CQ calls.

Your contribution to the image of the Amateur Radio Service is important. You can help maintain a self-disciplined, self-policing service that has high standards.

Distress Calls

If you require immediate emergency help, call MAYDAY. Use whatever frequency offers the best chance of getting a useful answer. "MAYDAY" is from the French "m'aider" (help me). On CW, use $\overline{SOS}$ to call for help. Repeat this call a few times, and pause for any station to answer. Identify the transmission with your call sign. Stations that hear your call sign will realize that the $\overline{SOS}$ is legitimate. Repeat this procedure for as long as possible, or until you receive an answer. Be ready to supply the following information to the stations responding to an $\overline{SOS}$ or MAYDAY:
* *The location of the emergency*, with enough detail to permit rescuers to locate it without difficulty.
* *The nature of the distress.*
* *The type of assistance required* (medical, evacuation, food, clothing or other aid.)
* *Any other information* to help locate the emergency area.

[Now turn to Chapter 10 and study questions 3AB-2-1.7, 3AB-2-2.1, 3AB-2-2.2, 3AB-3.1 through 3AB-3.3 and questions 3AB-4.1 and 3AB-4.2. Review this section if you have difficulty with any of these questions.]

EMERGENCY COMMUNICATIONS

RACES

Emergency communications are a major part of the radio amateur's purpose. Section 97.1, The Basis and Purpose of Amateur Radio, includes emergency communications. Government agencies are realizing the potential service of hams familiar with emergency communications techniques. FEMA, the Federal Emergency Management Agency, is responsible for the Radio Amateur Civil Emergency Service, RACES.

RACES is part of the Amateur Radio Service and provides radio communications for civil defense purposes. It is active *only* during periods of local, regional or national civil emergencies.

You must be registered with the responsible civil defense organization to operate as a RACES station. RACES stations may not communicate with amateurs not operating in a RACES capacity. Restrictions do not apply when stations are operating in a non-RACES amateur capacity, such as the ARES, the Amateur Radio Emergency Service. Only civil-preparedness communications can be transmitted during RACES operation. These are defined in Section 97.161 of the FCC regulations. Rules permit tests and drills for a maximum of one hour per week. All test and drill messages must be clearly identified as such.

The Amateur Radio Emergency Service (ARES) is an emergency preparedness group of about 25,000 amateurs. They have signed up voluntarily to keep Amateur Radio in the forefront of public service operating. ARES members provide communications at the city or county level.

[Now study questions 3AB-5-1.1, 3AB-5-1.2 and 3AB-5-2.1 in Chapter 10. Review as needed.]

Tactical Traffic

Tactical traffic is first-response communications in an emergency involving a few people in a small area. Tactical traffic is unformatted and seldom written. It may be urgent instructions or requests such as "Send an ambulance," or "Who has the medical supplies?" Tactical communications usually use 2-meter repeater net frequencies or the 146.52 MHz simplex calling frequency. Compatible mobile, portable and fixed-station equipment is plentiful and popular for these frequencies. Tactical traffic is particularly important in localized communications when working with government and law-enforcement agencies. In some relief activities, tactical traffic takes the form of operational or administrative traffic. Use the 12-hour local-time system for times and dates when working with relief agencies. Most may not understand the 24-hour system or Coordinated Universal Time (UTC).

Another way to make tactical traffic clear is to use tactical call signs. Tactical call signs describe a function, location or agency. Their use prevents confusion among listeners or agencies who are monitoring. When operators change shifts or locations, the set of tactical calls remains the same. Amateurs may use tactical call signs such as "parade headquarters," "finish line," "Red Cross" or "Net Control." This procedure promotes efficiency and coordination in public-service communication activities. Tactical call signs do not fulfill the identification requirements of Section 97.84 of the FCC rules. Amateurs must identify their station operation with their FCC-assigned call sign. Identify at the end of operation and at intervals not to exceed 10 minutes during operation.

Health-and-Welfare Traffic

There can be a large amount of radio traffic to handle during a disaster. Phone lines still in working order are often overloaded. They should be reserved for emergency use by those people in peril. Shortly after a major disaster, *Emergency* messages leave the disaster area. These have life-and-death urgency or are for medical help and critical supplies. Handle them first. Next is *Priority* traffic. These are emergency-related messages, but not of the utmost urgency. Then, *Welfare* traffic from evacuees or the injured flows out of the area. This results in timely advisories to those waiting outside the disaster area.

Hams inside disaster areas cannot immediately find out about someone's Aunt Irene when their own lives are threatened. They are busy handling emergency messages. Leave them to the tasks of surviving until things quiet down. Later, amateurs patiently waiting outside the disaster area can originate their Welfare messages. These messages are from concerned friends and relatives of the possible victims. They can ask about property damage within the disaster-affected area. Thus, the incoming traffic begins.

Avoid actively soliciting Welfare traffic for an emergency area. It can severely overload an already busy system when guaranteed delivery and reply is uncertain. Incoming Welfare traffic should be handled only after all emergency and priority traffic has been cleared. Welfare inquiries can take time to discover their hard-to-find answers. An advisory to the inquirer uses even more time. Meanwhile, some questions might already be answered through restored normal communications circuits.

Emergency Equipment

Amateur Radio operators provide a pool of experienced operators in time of national or community need. When cities, towns, counties, states or the federal government call for emergency communications, amateurs answer. They press their mobile and portable radio equipment into service where needed. They use alternatives to commercial power. Generators, car batteries, windmills or solar energy provide power for equipment during an emergency. Add some wire in the form of a dipole and you are ready to go on the air at any time and any place.

[Before you go on to Chapter 3, turn to Chapter 10 and study questions 3AB-6-1.1 through 3AB-6-3.2. Review this section if you have difficulty with any of these questions.]

Key Words

Backscatter—A small amount of signal that is reflected from the earth's surface after traveling through the ionosphere. The reflected signals may go back into the ionosphere along several paths and be refracted to earth again. Backscatter can help provide communications into a station's skip zone.

Critical angle—If radio waves leave an antenna at an angle greater than the critical angle for that frequency, they will pass through the ionosphere instead of returning to earth.

Critical frequency—The highest frequency at which a vertically incident radio wave will return from the ionosphere. Above the critical frequency, radio signals pass through the ionosphere instead of returning to the earth.

D layer—The lowest layer of the ionosphere. The D layer contributes very little to short-wave radio propagation. It acts mainly to absorb energy from radio waves as they pass through it. This absorption has a significant effect on signals below about 7.5 MHz during daylight.

Direct wave—A radio wave traveling in a straight line between the transmitting antenna and the receiving antenna.

Duct—A radio waveguide formed when a temperature inversion traps radio waves within a restricted layer of the atmosphere.

E layer—The second lowest ionospheric layer, the E layer exists only during the day, and under certain conditions may refract radio waves enough to return them to earth.

F layer—A combination of the two highest ionospheric layers, the F1 and F2 layers. The F layer refracts radio waves and returns them to earth. The height of the F layer varies greatly depending on the time of day, season of the year and amount of sunspot activity.

Guided propagation—Radio propagation by means of ducts.

Ionosphere—A region in the atmosphere about 30 to 260 miles above the earth. The ionosphere is made up of charged particles, or ions.

Maximum usable frequency (MUF)—The highest frequency that allows a radio wave to reach a desired destination.

Propagation—The means by which radio waves travel from one place to another.

Radio-path horizon—The point where radio waves are returned by tropospheric bending. The radio-path horizon is 15 percent farther away than the geometric horizon.

Reflected wave—A radio wave whose direction is changed when it bounces off some object in its path.

Refract—To bend. Electromagnetic energy is refracted when it passes through a boundary between different types of material. Light is refracted as it travels from air into water or from water into air.

Skip zone—A region between the farthest reach of ground-wave communications and the closest range of skip propagation.

Sky waves—Radio waves that travel from an antenna upward to the ionosphere, where they either pass through the ionosphere into space or are refracted back to earth.

Solar flux index—A measure of solar activity. The solar flux index is a measure of the radio noise on 2800 MHz.

Space wave—A radio wave arriving at the receiving antenna made up of a direct wave and one or more reflected waves.

Sunspots—Dark blotches that appear on the surface of the sun.

Temperature inversion—A condition in the atmosphere in which a region of cool air is trapped beneath warmer air.

Troposphere—The region in the earth's atmosphere just above the surface of the earth and below the ionosphere.

Tropospheric bending—When radio waves are bent in the troposphere, they return to earth approximately 15 percent farther away than the geometric horizon.

True or **Geometric horizon**—The most distant point one can see by line of sight.

Virtual height—The height that radio waves appear to be reflected from when they are returned to earth by refraction in the ionosphere.

Chapter 3

Radio-Wave Propagation

R adio amateurs have studied radio **propagation** since the early days of radio. Amateurs want to know what influences radio waves as they travel. Antenna height, type of antenna and frequency of operation affect propagation. Terrain, weather and the height and density of the **ionosphere** all affect how radio waves travel. Understanding the nature of radio waves is useful. Knowledge of how their behavior is affected by the medium they travel in is important.

IONOSPHERIC PROPAGATION

Nearly all amateur communication below 30 MHz is by **sky waves.** This type of wave leaves the transmitting antenna and travels from the earth's surface at an angle. It would go into space if it was not bent back to earth. As the radio wave travels outward, it enters a region of ionized particles. This region, the ionosphere, begins about 30 miles up and extends to about 260 miles. The ionosphere **refracts**, or bends, radio waves. At some frequencies the radio waves are refracted enough so they return to earth. See Figure 3-1. These signals often return far from the originating station.

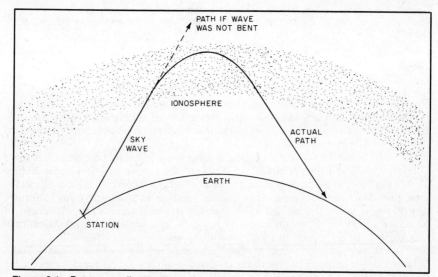

Figure 3-1—Because radio waves are bent in the ionosphere, they return to earth far from their origin. Without refraction in the ionosphere, radio waves would pass into space.

The earth's upper atmosphere consists mainly of oxygen and nitrogen. There are traces of hydrogen, helium and several other gases. The atoms making up these gases are electrically neutral. They have no charge and exhibit no electrical force outside their own structure. The gas atoms absorb ultraviolet radiation from the sun, which knocks electrons out of the atom. In this process, the atoms become positively charged. A positively charged atom is an ion. The process by which ions are formed is ionization. Several ionized layers appear at different heights in the atmosphere. Each layer has a central region where the ionization is greatest. The intensity of the ionization decreases above and below this central region in each layer. See Figure 3-2.

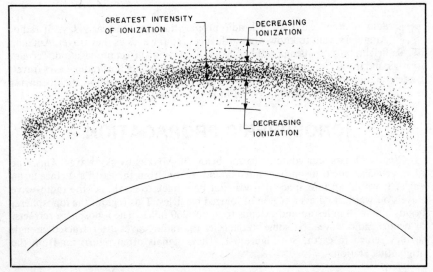

Figure 3-2—A cross section of a layer of the ionosphere. The intensity of the ionozation is greatest in the central region and decreases above and below the central region.

The Ionosphere: A Closer Look

The ionosphere consists of several layers of charged particles. These layers have been given letter designations, as shown in Figure 3-3. Scientists started with the letter D just in case there were any undiscovered lower layers. None have been found, so there is no A,B or C layer.

The **D Layer:** The lowest layer of the ionosphere affecting propagation is the D layer. This layer is in a relatively dense part of the atmosphere. It is about 30 to 55 miles above the earth. When the atoms in this layer absorb sunlight and form ions, they don't last very long. They quickly combine with free electrons to form neutral atoms again. The amount of ionization in this layer varies widely. It depends on how much sunlight hits the layer. At noon, D-layer ionization is maximum or very close to it. By sunset, this ionization disappears.

The D layer is ineffective in bending high-frequency signals back to earth. The D layer's major effect is to absorb energy from radio waves. As radio waves pass through the ionosphere, they give up energy. This sets some of the ionized particles into motion. Effects of absorption on lower frequencies are greater than on higher frequencies. Absorption also increases proportionally with the amount of ionization.

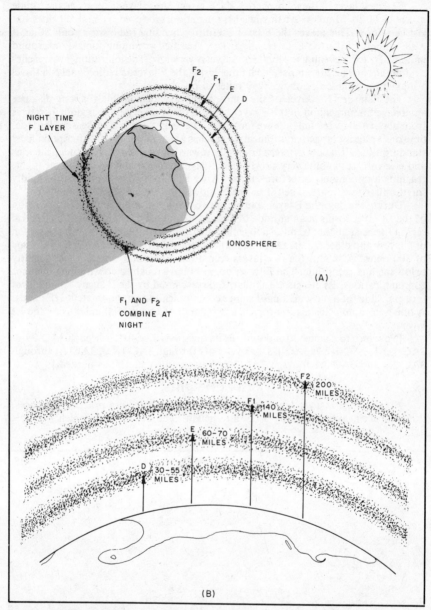

Figure 3-3—The ionosphere consists of several layers of ionized particles at different heights above the earth. At night, the D and E layers disappear and the F1 and F2 layers combine to form a single F layer.

The more ionization, the more energy the radio waves lose passing through the ionosphere. Absorption is most pronounced at midday. It is responsible for the short daytime communications ranges on the lower amateur frequencies (160, 80 and 40 meters).

The next layer of the ionosphere is the **E layer**. The E layer appears at an altitude of about 60 to 70 miles. At this height, ionization produced by sunlight does not last very long. This makes the E layer useful for bending radio waves only when it is in sunlight. Like the D layer, the E layer reaches maximum ionization around midday. By early evening the ionization level is very low. The ionization level reaches a minimum just before sunrise, local time. Using the E layer, a radio signal can travel a maximum distance of about 1250 miles in one hop.

The **F layer**: The layer of the ionosphere most responsible for long-distance amateur communication is the F layer. This layer is a very large region. It ranges from about 100 to 260 miles above the earth. The height depends on season, latitude, time of day and solar activity. Ionization reaches a maximum shortly after noon local standard time. It tapers off very gradually toward sunset. At this altitude, the ions and electrons recombine very slowly. The F layer remains ionized during the night, reaching a minimum just before sunrise. After sunrise, ionization increases rapidly for the first few hours. Then it increases slowly to its noontime maximum.

During the day, the F layer splits into two parts, F1 and F2. The central region of the F1 layer forms at an altitude of about 140 miles. For the F2 layer, the central region forms at about 200 miles above the earth. These altitudes vary with the season of the year and other factors. At noon in the summer the F2 layer can reach an altitude of 300 miles. At night, the two layers recombine to form a single F layer slightly below the higher altitude. The F1 layer does not have much to do with long-distance communications. Its effects are similar to those caused by the E layer. The F2 layer is responsible for almost all long-distance communication on the amateur HF bands. A one-hop radio transmission travels a maximum of about 2500 miles using the F2 layer.

[Now turn to chapter 10 and study exam questions 3AC-1-1.1 through 3AC-1-1.3, 3AC-1-2.1, 3AC-1-2.2, 3AC-1-3.1, 3AC-1-4.1 through 3AC-1-4.3, 3AC-2.1 through 3AC-2.4 and 3AC-3.1 through 3AC-3.4. Review this material as needed.]

Virtual Height

An ionospheric layer is a region of considerable depth. For practical purposes, think of each layer as having a definite height. The height from which a simple

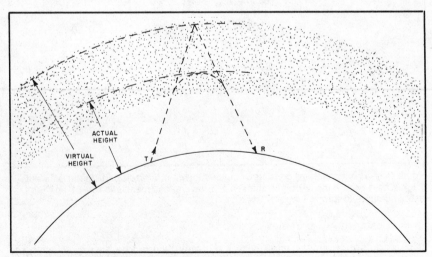

Figure 3-4—the virtual height of a layer in the ionosphere is the height at which a simple reflection would return the wave to the same point as the gradual bending that actually takes place.

reflection from the layer would give the same effects (observed from the ground) as the effects of the gradual bending that actually takes place is called the **virtual height**. See figure 3-4.

The virtual height of an ionospheric layer for various frequencies is determined with an ionosonde. An ionosonde is a variable-frequency transmitter and receiver. It directs radio energy vertically and measures the time required for the signal to make a round-trip. As the frequency increases, there is a point where no energy will return. The highest frequency at which vertically incident waves return to earth is the **critical frequency**.

Radiation Angle and Skip Distance

A radio wave at a low angle above the horizon requires less refraction to bring the wave back to earth. This is why we want antennas with low radiation angles to work DX.

Figure 3-5 illustrates some of the effects of radiation angle. The high-angle waves are bent only slightly in the ionosphere and pass through it. The wave at the somewhat lower angle is just able to return. In daylight it might be returned from the E layer. The point of return for high angle waves is relatively close to the transmitting station. The lowest-angle wave returns farther away, to point B.

The highest radiation angle that will return a radio wave to earth under specific ionospheric conditions is the **critical angle**. Waves meeting the ionosphere at greater than the critical angle will pass through the ionosphere into space.

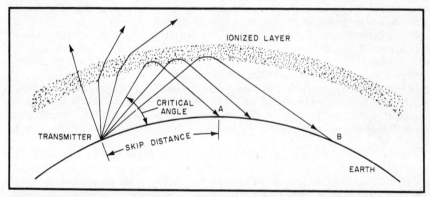

Figure 3-5—Ionospheric propagation. The waves at left leave the transmitter above the critical angle—they are refracted in the ionosphere, but not enough to return to earth. The wave at the critical angle will return to earth. The lowest-angle wave will return to earth farther away than the wave at the critical angle. This explains the emphasis on low radiation angles for DX work.

Maximum Usable Frequency

There is little doubt that the critical frequency is important to amateur communication. More interesting to radio amateurs is the frequency *range* over which communication can be carried via the ionosphere. Most amateurs want to know the **maximum usable frequency** (abbreviated MUF). The MUF applies to a particular direction and distance at the time of day when you want to communicate. The MUF is the highest frequency that allows a radio wave to reach the desired destination using E-or F-layer propagation. The MUF is affected by the amount of ultraviolet and other types of radiation received from the sun. The MUF is subject to seasonal variations as well as changes throughout the day.

Why is the MUF so important? If we know the MUF, we can make accurate predictions about which bands give the best chance for communication over a particular path. Predictions concerning amateur bands affording the best chance for communication via particular paths can be made. For example, we wish to contact a distant station. We know that the MUF for that path is 17 MHz at the time we wish to make our contact. The closest amateur band below that frequency is the 20-meter, or 14-MHz, band. The 20-meter band offers our best chance to contact that long distance location, then. There is no single MUF for a given location at any one time. The MUF will vary depending on the direction and distance to the station we wish to contact. Understanding and using the MUF is one way you can change your study of propagation from guesswork into a science.

[Now turn to Chapter 10 and study examination questions 3AC-4.1 through 3AC-4.3. Review this section as needed.]

Solar Activity and Radio-Wave Propagation

It should be clear that the sun plays the major role in the ionization of different layers in the ionosphere. The sun is the dominant factor in sky-wave communication. To communicate effectively over long distances, you must understand how solar conditions will affect your radio signals.

You know about the day-to-day sun-related cycles on earth, such as the time of day and the season of the year. Conditions affecting radio communication vary with these same cycles. Long-term and short-term solar cycles also influence propagation in ways that are not so obvious. The condition of the sun at any given moment has a very large effect on long-distance radio communication. The sun is what makes propagation prediction an inexact science.

Man's interest in the sun is older than recorded history. **Sunspots** are dark regions that appear on the surface of the sun. They were observed and described thousands of years ago. Observers noted that the number of sunspots increased and decreased in cycles. The solar observatory in Zurich, Switzerland has been recording solar data on a regular basis since 1749. The solar cycle that began in 1755 was designated cycle 1. The low sunspot numbers marking the end of cycle 21 and the beginning of cycle 22 occurred in September of 1986. Sunspot cycles also vary a great deal. For example, cycle 19 peaked with a smoothed mean sunspot number of over 200. Cycle 14 peaked at only 60.

Maximum ionization occurs during a sunspot cycle peak. High sunspot numbers usually mean good worldwide radio communications. During sunspot cycle peaks, the 20-meter amateur band is open to distant parts of the world almost continuously. The cycle 19 peak in 1957 and 1958 was responsible for the best propagation conditions in the history of radio. Sunspot cycles do not follow consistent patterns. There can be periods of high activity that seem to come from nowhere during periods of low sunspot activity.

There is an important clue to anticipating variations in solar radiation levels. Radio propagation changes resulting from these variations can be predicted. This clue is the time it takes the sun to rotate on its axis, approximately 27 days. Active areas capable of influencing propagation can recur at four-week intervals. These active areas may last for four or five solar rotations. High MUF and good propagation for several days indicates similar conditions may develop approximately 27 days later.

Another useful indication of solar activity is solar flux, or radio energy coming from the sun. Increased solar activity produces higher levels of solar energy. More solar energy produces greater ionization in the ionosphere. Scientists use sophisticated receiving equipment and large antennas pointed at the sun to measure solar flux. A number called the **solar-flux index** is given to represent the amount of solar flux. The solar-flux index is gradually replacing the sunspot number as a means of predicting radio-wave propagation.

The solar-flux measurement is taken at 1700 UTC daily in Ottawa, Canada, on 2800 MHz (10.7 centimeters). The information is then transmitted by the National Institute of Science and Technology station WWV in Fort Collins, Colorado. Both the sunspot number and the solar flux measurements tell us similar things about solar activity. To get the sunspot number, the sun must be visible. The solar-flux measurement may be taken under any weather conditions.

Using the Solar Flux

We can use the solar flux numbers to make general daily predictions about band conditions. Figure 3-6 shows a relationship between average sunspot numbers and the 2800-MHz solar-flux measurement. The solar flux varies directly with the activity on the sun. Values range from around 60 to 250 or so. Flux values in the 60s and 70s mean fair to poor propagation conditions on the 14-MHz (20-meter) band and higher frequencies. Values from 90 to 110 or so indicate good conditions up to about 24 MHz (12 meters). Solar flux values over 120 indicate very good conditions on 28 MHz (10 meters) and even up to 50 MHz (6 meters) at times. On 6-meters, the flux values must be in the 200s for reliable long-range communications. The higher the frequency, the higher the flux value required for good propagation.

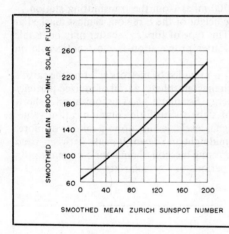

Figure 3-6—The relationship between smoothed Zurich sunspot numbers and 2800-MHz solar flux values.

The Scatter Modes

All electromagnetic-wave propagation is subject to scattering influences. These alter idealized patterns to a great degree. The earth's atmosphere, ionospheric layers and any objects in the path of radio signals scatter the energy. Understanding how scattering takes place helps us use this propagation mode to our advantage.

Forward Scatter

There is an area between the outer limit of ground-wave propagation and the point where the first signals return from the ionosphere. We call this area the **skip zone**. This zone is shown in Figure 3-7. The skip zone is often described as if communications between stations in each other's skip zone were impossible. Actually, some of the transmitted signal is scattered in the atmosphere, so the signal can be heard over much of the skip zone. You usually need a very sensitive receiver and good operating techniques to hear these signals.

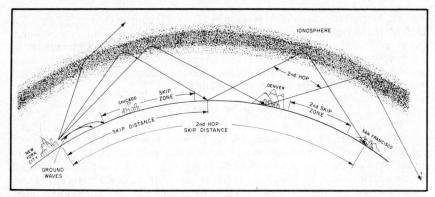

Figure 3-7—There is an area between the farthest reaches of ground-wave propagation and the closest return of sky waves from the ionosphere. This region is known as the skip zone.

The **troposphere** is a region of the atmosphere below the ionosphere. VHF "tropo" scatter is usable out to about 500 miles from the transmitting station.

Ionospheric scatter, mostly from the height of the E region, is most marked at frequencies up to about 60 or 70 MHz. This type of forward scatter may be usable out to about 1200 miles. Ionospheric scatter propagation is most noticeable on frequencies above the MUF.

Another means of ionospheric scatter is provided by meteors entering the earth's atmosphere. As the meteor passes through the ionosphere, a trail of ionized particles forms. These particles can scatter radio energy. See figure 3-8. This ionization is short-lived and can show up as short bursts of signals with little communication value. There may be longer periods of usable signal levels, lasting up to a minute or more. Meteor scatter is most common between midnight and dawn. It peaks between 5 and 7 AM local time. Meteor scatter is an interesting method of amateur communication at 21 MHz or higher, especially during periods of low solar activity.

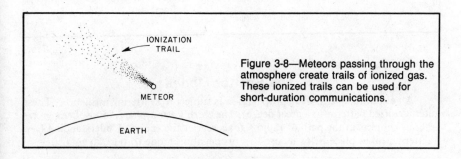

Figure 3-8—Meteors passing through the atmosphere create trails of ionized gas. These ionized trails can be used for short-duration communications.

Backscatter

You can observe a complex form of scatter when you are working very near the MUF. The transmitted wave is refracted back to earth at some distant point. This may be an ocean area or land mass. A portion of the transmitted signal reflects back into the ionosphere. Some of this signal comes toward the transmitting station. The reflected wave helps fill in the skip zone, as shown in Figure 3-9.

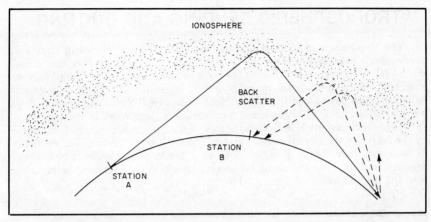

Figure 3-9—When radio waves strike the ground after passing through the ionosphere, some of the signal may reflect back into the ionosphere. If some of the signal is reflected from the ionosphere toward the transmitting station, the energy may be scattered back into the skip zone.

Backscatter signals are generally weak and subject to distortion, because the signal may arrive at the receiver from many different directions. Backscatter is usable from just beyond the local range out to several hundred miles. Under ideal conditions, backscatter is possible over 3000 miles or more. The term "sidescatter" is more descriptive of what probably happens on such long paths, however.

[Now study exam questions 3AC-5.1 and 3AC-5.2 in Chapter 10. Review the material in this section as needed.]

LINE-OF-SIGHT PROPAGATION

In the VHF and UHF range (above 144 MHz), most propagation is by **space wave**. See Figure 3-10. Space waves are made up of a **direct wave**, and usually one or more **reflected waves**. The direct wave and reflected wave arrive at the receiving station slightly out of phase. They may cancel or reinforce each other to varying degrees. This depends on the actual path of the reflected wave(s). Effects are most pronounced when either, or both, stations are mobile. When one or both stations are moving, the paths are constantly changing. The waves may alternately reinforce and cancel each other. This causes a rapid fluttering sound (called "picket-fencing").

[At this point you should turn to Chapter 10 and study questions 3AC-6.1 and 3AC-6.2. Review any material necessary.]

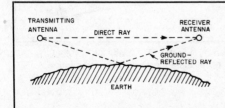

Figure 3-10—Line-of-sight propagation is accomplished by means of the space wave, a combination of a direct ray and one or more reflected rays.

TROPOSPHERIC BENDING AND DUCTING

The **troposphere** consists of atmospheric layers close to the earth's surface. **Tropospheric bending** is evident over a wide range of frequencies. It is most useful in the VHF/UHF region, especially at 144 MHz and above. Instead of gradual changes in air temperature, pressure and humidity, distinct layers may form in the troposphere. Adjacent layers having significantly different densities will bend radio waves passing between layers. Light is bent when it passes from air into water in the same way.

The **true** or **geometric horizon** is the most distant point one can see. It is limited by the height of the observer above ground. Slight bending of radio waves occurs in the troposphere. This will cause signals to return to earth somewhat beyond the geometric horizon. This **radio-path horizon** is generally about 15 percent farther away than the true horizon. See Figure 3-11.

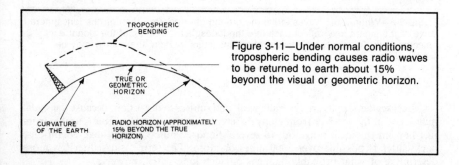

Figure 3-11—Under normal conditions, tropospheric bending causes radio waves to be returned to earth about 15% beyond the visual or geometric horizon.

Under normal conditions, the temperature of the air gradually decreases with increasing height above ground. Under certain conditions, however, a mass of warm air may overrun cold air. Then there is an area where cold air is covered by warm air. This is a **temperature inversion**. Radio waves can be trapped below the warm air mass. They can travel great distances with little loss. The area between the earth and the warm air mass is known as a **duct**. See figure 3-12.

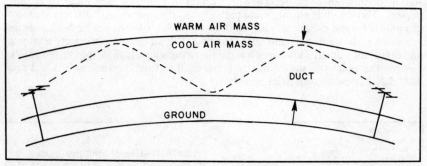

Figure 3-12—When a cool air mass is overrun by a mass of warmer air, a "duct" is formed, allowing VHF radio signals to travel great distances with little attenuation.

The term for radio-wave propagation through a duct is **guided propagation,** sometimes called tropospheric ducting. The manner of guiding waves through the duct is very similar to microwaves traveling in waveguides. As with tropospheric bending, we find ducting primarily at VHF frequencies, 144 MHz and higher. Ducts usually form over water, though they can form over land as well. The lowest usable frequency for duct propagation depends on two factors. One is the depth of the duct. The other is the amount the refractive index changes at the air-mass boundary.

[Before going on to Chapter 4, turn to Chapter 10 and study questions 3AC-7.1 through 3AC-7.6. Review this section as needed.]

Key Words

Balun—Short for BALanced to UNbalanced. The balun is used to transform between an unbalanced feed line and a balanced antenna.

Dummy load (dummy antenna)—A resistor that acts as a load for a transmitter, dissipating the output power without radiating a signal. A dummy load is used when testing transmitters.

Field-effect transistor volt-ohm milliammeter (FET VOM)—A multiple-range meter used to measure voltage, current and resistance. The meter circuit uses an FET amplifier to provide a high input impedance for more accurate readings.

Marker generator—An RF signal generator that produces signals at known frequency intervals. The marker generator can be used to calibrate receiver and transmitter frequency readouts.

Multimeter—An electronic test instrument used to make basic measurements of current and voltage in a circuit. This term is used to describe all meters capable of making different types of measurements, such as the VOM, VTVM and FET VOM.

Reflectometer—A test instrument used to indicate standing wave ratio (SWR) by measuring the forward power (power from the transmitter) and reflected power (power returned from the antenna system).

Signal generator—A test instrument that produces a stable low-level radio-frequency signal. The signal can be set to a specific frequency and used to troubleshoot RF equipment.

S meter—A meter in a receiver that shows the relative strength of a received signal.

Vacuum-tube voltmeter (VTVM)—A multimeter using a vacuum tube in its input circuit. The VTVM does not load a circuit as much as a VOM, and can be used to measure small voltages in low-impedance circuits. A VTVM generally requires 120 V ac power.

Volt-ohm-milliammeter (VOM)—A test instrument used to measure voltage, resistance and current. A VOM usually has a RANGE switch so the meter can be used to measure a wide range of inputs.

Wattmeter—A test instrument used to measure the power output (in watts) of a transmitter.

Chapter 4

Amateur Radio Practice

It's easy to form good operating habits in Amateur Radio. You will need to understand the fundamentals of radio. Then you can maintain your station to operate within FCC rules and regulations. Also, if you're like most amateurs, you'll want to put out the best possible signal. Know the principles behind the operation of your station. Correctly adjust your equipment and measure its performance. Take advantage of available test equipment. You should have no problem keeping your station "up to par."

ELECTRICAL WIRING SAFETY

Your station equipment makes use of ac line voltage. This voltage can be dangerous. The equipment also generates additional potentially lethal voltages of its own. You should be familiar with some basic precautions. Your own safety and that of others depends on it.

Power-Line Connections

In most residential systems, three wires come in from outside to the distribution board. Older systems may use only two wires. In the three-wire system, the voltage between the two "hot" wires is normally 240. The third wire is neutral and is grounded. Half of the total voltage appears between each of the hot wires and neutral, as shown in Figure 4-1. Lights, appliances and 120-V outlets are divided as evenly as possible

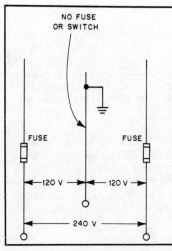

Figure 4-1—In home electrical systems, two wires carrying 240 volts are split between the house circuits carrying 120 volts each. Some heavy appliances use the full 240 V. As shown, a fuse should be placed in each hot wire, but no fuse or switch should be placed in the neutral wire.

between the two sides of this circuit. Half of the load connects between one hot wire and the neutral. The other half of the load connects between the other hot wire and neutral.

Heavy appliances, such as electric stoves and most high-power amateur amplifiers, are designed for 240-V operation. Connect these across the two hot wires. *Both* ungrounded wires should be fused. The neutral wire should *never* have a fuse or switch in it.

Four-conductor appliance power cords should have the black and red wires connected to the hot wires with fuses. The white and green (or bare) wires should not have a fuse or switch. Opening a switch in the neutral wire does not disconnect the equipment from the household voltage. It places the equipment on one side of the 240-V line in series with anything across the other side. See Figure 4-2. When the neutral wire is open, the voltage will divide between the two loads. The division will be in proportion to each load resistance. The voltage will go above normal on the side with the larger load resistance. On the other side, the voltage will be lower than normal. This will probably destroy the appliance. If both loads happen to be equal, the voltages will divide normally.

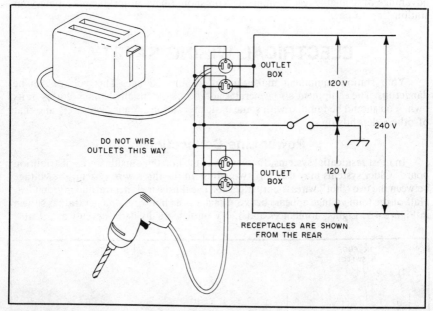

Figure 4-2—Opening the neutral wire leaves the equipment on one side of the 240-V line in series with anything connected on the other side. This is extremely dangerous!

High-power amateur amplifiers use the full 240 V. It only takes half as much current to supply the same power as with a 120-V line. The power-supply circuit efficiency improves with the higher voltage and lower current. A separate 240-V circuit for the amplifier ensures that you stay within the capacity of the 120-V circuits in your shack. An amplifier drawing more current than the house wiring can handle could cause the house lights to dim. This is because the heavy current causes the power-line voltage to drop.

Three-Wire 120-V Power Cords

State and national electrical-safety codes require three-wire power cords on many 120-V tools and appliances. Two of the conductors (the "hot" and "neutral" wires) power the device. The third conductor (the safety ground wire) connects to the metal frame of the device. See Figure 4-3. The "hot" wire is usually black or red. The "neutral" wire is white. The frame/ground wire is green or sometimes bare.

Let's look at the power cord connections for a transmitter power supply. The power supply will have a transformer in it. When you attach the power cord, the black wire will attach to one end of the fuse. The white wire will go to the side of the transformer primary winding without a fuse. The green (or bare) wire will attach to the chassis.

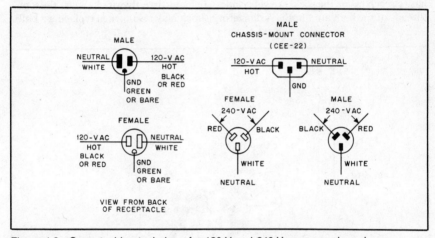

Figure 4-3—Correct wiring technique for 120-V and 240-V power cords and receptacles. The white wire is neutral, the green wire is ground and the black or red wire is the hot lead. Notice that the receptacles are shown as viewed from the back, or wiring side.

When plugged into a properly wired mating receptacle, a three-contact plug connects the third conductor to an earth ground. This grounds the appliance chassis or frame and prevents the possibility of electric shock. A defective power cord that shorts to the case of the appliance will simply blow a fuse. Without the ground connection, the case could carry the full line voltage, presenting a severe shock hazard. All commercially manufactured electronic test equipment and most ac-operated amateur equipment uses these three-wire cords. Adapters are available for use where older electrical installations do not have mating receptacles. The lug of the green wire from the adapter must be attached under the cover-plate screw. The outlet (and outlet box) must be grounded for this to be effective. Power wires coming into the electric box inside a flexible metal covering provide grounding through the metal covering. The common name for this type of wire is armored cable.

A "polarized" two wire plug has one blade that is wider than the other. The mating receptacle will accept the plug only one way. This ensures that the hot and neutral wires in the appliance connect to the appropriate wires in the house electrical system. Consider what happens without this polarized plug and receptacle. The power switch in the equipment will be in the hot wire when the plug is inserted one way. It will be in the neutral wire when inserted the other way. This can present a dangerous condition. It is possible for the equipment to be "hot" even with the switch off. With the switch in the neutral line, the hot line may be connected to the equipment chassis. An unsuspecting operator could form a path to ground by touching the case, and might receive a nasty shock!

Wiring an outlet or lamp socket properly is important. See Figure 4-4. The black (hot) wire should be connected to the brass terminal on the lamp socket or outlet. The white (neutral) wire should be connected to the white or silver-colored terminal. This will ensure that the proper blade of the plug connects to the hot wire. This practice is especially important when wiring lamp sockets. The brass screw of the socket connects to the center pin in the socket. With this pin as the "hot" connection, there is much less shock hazard. Anyone unscrewing a bulb from a correctly wired socket has to reach inside the socket to get a shock. In an incorrectly wired socket, the screw threads of the bulb are "hot". A dangerous shock may result when replacing a bulb.

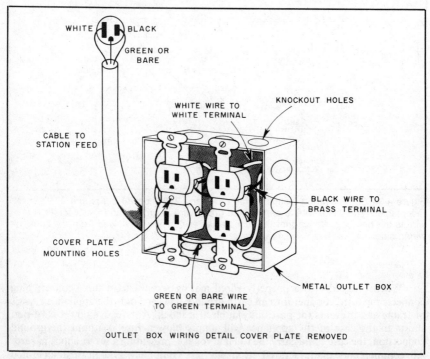

Figure 4-4—The correct way to wire a receptacle box. Connect the white wire to the white or silver terminal, and the black wire to the brass-colored terminal. This ensures that the prongs on the mating plug will be connected properly.

Current Capacity

There is another factor to take into account when you are wiring an electrical circuit. It is the current-handling capability of the wire. Table 4-1 shows the current-handling capability of some common wire sizes. The table shows that number 14 wire could be used for a circuit carrying 15 A. You must use number 12 (or larger) for a circuit carrying 20 A.

Table 4-1

Current-Carrying Capability of Some Common Wire Sizes

Wire Size (AWG)	Continuous-Duty Current*
8	46 A
10	33 A
12	23 A
14	17 A
16	13 A
18	10 A
20	7.5 A
22	5 A

*wires or cables in conduits or bundles

To remain safe, don't overload the ac circuits in your home. The circuit breaker or fuse rating is the maximum load for the line at any one time. Simple mathematics can be used to calculate the current your ham radio equipment will draw. Most equipment has the power requirements printed on the back. If not, the owner's manual should contain such information. Current calculations should include any other household appliances on the same line, including lights! If you put a larger fuse in the circuit, too much current could be drawn. The wires would become hot and a fire could result.

[Now turn to Chapter 10 and study questions 3AD-1-1.1 through 3AD-1-1.4. Review the material in this section if you have any difficulty with any of these questions.]

Power-Supply Safety

Safety must always receive careful consideration during the design and construction of any power supply. Power supplies can produce potentially lethal currents and voltages. Be careful to guard against accidental exposure to these currents and voltages. Use electrical tape, insulated tubing (spaghetti) or heat-shrink tubing to cover exposed wires. This includes component leads, component solder terminals and tie-down points. Whenever possible, connectors used to mate the power supply to the outside world should be of an insulated type. They should be designed to prevent accidental contact with the voltages present. AC power to the supply should be controlled by a clearly labeled front-panel switch. That way it can be seen and reached easily in an emergency.

All dangerous voltages in equipment should be made inaccessible. A good way to ensure this is to enclose all equipment in metal cabinets. That way no "hot" spots can be reached. Don't forget any component shafts that might protrude through the front panel. If a control shaft is hot, protect yourself from accidental contact by using an insulated shaft extension or insulated knob.

Each metal enclosure should be connected to a good earth ground, such as a ground rod. Then, if a failure occurs inside a piece of equipment, the metal case will never present a shock hazard. The fuse will blow instead.

It's also a good idea to make it impossible for anyone to energize your equipment when you're not present. A key-operated ac-mains switch that controls all power to your station is a good way to accomplish this. Mount your switch where it can be seen and reached easily in an emergency.

You should never underestimate the potential hazard when working with electricity. Table 4-2 shows some of the effects of electric current—as little as 100 mA can be lethal! As the saying goes, "It's volts that jolts, but it's mills that kills." Low-voltage power supplies may seem safe, but even battery-powered equipment should be treated with respect. Thirty volts is the minimum voltage considered dangerous to humans. These voltage and current ratings are only general guidelines. Automobile batteries are designed to provide very high current (as much as 200 A) for short periods when starting a car. This much current can kill you, even at 12 volts. You will feel pain if the shock current is in the range of 30 to 50 mA.

A few factors affect just how little voltage and current can be considered

Table 4-2

Effects of Electric Current Through the Body of an Average Person

(1 Second Contact)	Current Effect
1 mA	Just perceptible.
5 mA	Maximum harmless current.
10-20 mA	Lower limit for sustained muscular contractions.
30-50 mA	Pain
50 mA	Pain, possible fainting. "Can't let go" current.
100-300 mA	Normal heart rhythm disrupted. Electrocution if sustained current.
6 A	Sustained heart contraction. Burns if current density is high.

dangerous. One factor is skin resistance. The lower the resistance of the path, the more current that will pass through it. If you perspire heavily, you may get quite a bit more severe shock than if your skin is dry. Another factor is the path through the body to ground. As Figure 4-5 shows, the most dangerous path is from one hand to the other. This path passes directly through a person's heart. Even a very minimal current can cause heart failure and death. Current passing from one finger to another on the same hand will not have quite such a serious effect. For this reason, if you must troubleshoot a live circuit, keep one hand behind your back or in your pocket. If you do slip, the shock may not be as severe as if you were using both hands.

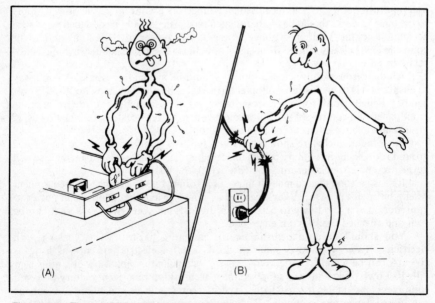

Figure 4-5—The path from the electrical source to ground affects how severe an electrical shock will be. The most dangerous path (from hand to hand directly through the heart) is shown at A. The path from one finger to the other shown at B is not quite so dangerous.

Bleeder Resistors

An important safety item in power-supply design is the bleeder resistor. When a power supply is turned off, the filter capacitors can store a charge for a long time. These charged capacitors present a shock hazard at the output terminals. A bleeder resistor connected across the filter capacitors will dissipate the charge stored in the capacitors when the supply is turned off. This will not affect normal operation of the supply because the bleeder resistor draws only a very small current.

[Now turn to Chapter 10 and study exam questions 3AD-1-2.1 through 3AD-1-2.3 and 3AD-1-3.1. Review any topics necessary before proceeding.]

USING TEST EQUIPMENT

The Voltmeter

The voltmeter is an instrument used to measure voltage. It is a basic meter movement with a resistor in series, as shown in Figure 4-6. The current multiplied by the resistance will be the voltage drop across the resistance. An instrument used this way is calibrated in terms of the voltage drop across the resistor to read voltage.

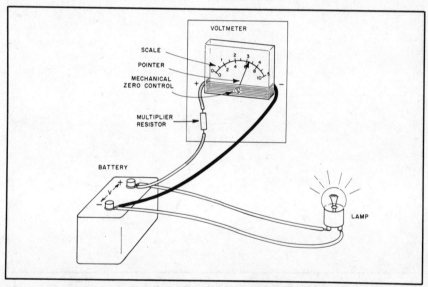

Figure 4-6—When you use a voltmeter to measure voltage, the meter must be connected in parallel with the voltage you want to measure.

A high value resistor provides a high impedance input for the meter. An ideal voltmeter would have an infinite input impedance. It would not affect the circuit under test in any way. Real-world voltmeters, however, have a finite value of input impedance. There is a possibility that the voltage you are measuring will change. This is because the meter adds some load to the circuit when you connect the voltmeter.

Use a voltmeter with a very high input impedance compared with the impedance of the circuit you are measuring. This prevents the voltmeter from drawing too much current from the circuit. Excessive current drawn from the circuit would significantly affect circuit operation. The input impedance of an ordinary voltmeter is about 20 kilohms per volt. Voltage measurements are made by placing the meter in parallel with the voltage to be measured.

The range of a meter can be extended to measure a wide range of voltages. This is accomplished by changing the resistor in series with it.

The Ammeter

The ammeter depends upon a current flowing through it to deflect the needle. The ammeter is placed in series with the circuit. That way, all the current flowing in the circuit must pass through the meter. Many times the meter cannot handle all the current. The range of the meter can be extended by placing resistors in parallel with the meter to provide a path for part of the current. These shunt resistors are shown in Figure 4-7. They are calculated so that the total circuit current can be read on the meter. Most ammeters have very low resistance, but low-impedance circuits require caution. There is a chance that the slight additional resistance of the series ammeter will disturb circuit operation.

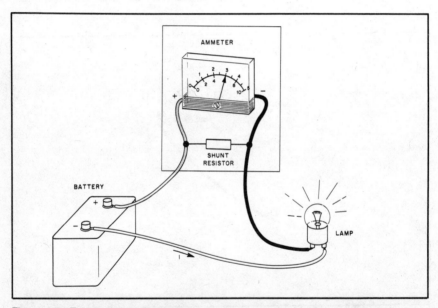

Figure 4-7—To measure current you must break the circuit at some point and connect the meter in series at the break. A shunt resistor expands the scale of the meter to measure higher currents than it could normally handle.

Multimeters

A **multimeter** is a piece of test equipment that most amateurs should know how to use. The simplest kind of multimeter is the **volt-ohm-milliammeter (VOM)**. VOMs use one basic meter movement for all functions. The movement requires a fixed amount of current (often 1 mA) for a full-scale reading. As shown in the two previous sections, resistors are connected in series or parallel to provide the proper meter reading. In a VOM a switch selects various ranges for voltage, resistance and current measurements. This switch places high-value dropping resistors in series with the meter movement for voltage measurement. It connects low-value shunt resistors in parallel with the movement for current measurement. These parallel and series resistors extend the range of the basic meter movement.

If you are going to purchase a VOM, buy one with the highest ohms-per-volt rating that you can find. Stay away from meters rated much under 20,000 ohms per volt if you can.

Measuring resistance with a meter involves placing the meter leads across the component or circuit you wish to measure. Make sure to select the proper resistance scale. The full-scale-reading multipliers vary from 1 to 1000 and higher. The scale is usually compressed on the higher end of the range. See Figure 4-8. For best accuracy, keep the reading in the lower-resistance half of the scale. On most meters this is the right-hand side. Thus, if you want to measure a resistance of about 5000 ohms, select the R × 1000 scale. Then the meter will indicate 5.

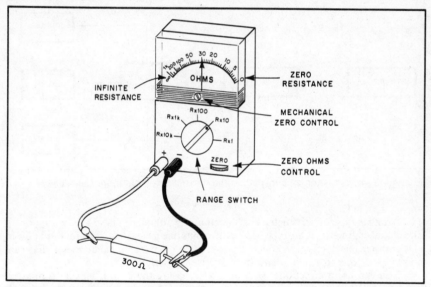

Figure 4-8—Resistance is measured across the component. For best accuracy the reading should be taken in the lower-resistance half of the scale.

A **vacuum-tube voltmeter (VTVM)** operates in the same manner as an ordinary VOM. There is one important difference. The meter in a VTVM is isolated from the circuit under test by a vacuum-tube dc amplifier. As a result, the only additional circuit loading is from the tube input impedance, which is very high. The standard VTVM input impedance is 11 MΩ. This is useful for measuring voltages in high-impedance circuits, such as vacuum-tube grid circuits and FET gate circuits.

Another type of meter is a **field-effect transistor volt-ohm-milliammeter (FET VOM)**. These instruments use a field-effect transistor (FET) to isolate the indicating meter from the circuit to be measured. FET VOMs have an input impedance of several megohms. They are the solid-state equivalent of a VTVM.

[Now turn to Chapter 10 and study questions 3AD-2-1.1, 3AD-2-2.1, 3AD-3-1.1, 3AD-3-2.1 and 3AD-4.1. Review this section if you have difficulty with any of those questions.]

Wattmeters

A **wattmeter** is a device in the transmission line to measure the power (in watts) coming out of a transmitter. Wattmeters are designed to operate at a certain line impedance, normally 50 ohms. Make sure that the feed line impedance is the same as the design impedance of the wattmeter. If impedances are different, any measurements will be inaccurate. For most accurate measurement, the wattmeter should be connected directly at the transmitter antenna jack as shown in Figure 4-9.

All wattmeters must contain some type of detection circuitry to enable you to measure power. The accuracy and upper frequency range of the wattmeter is limited by the detection device. Stray capacitance and coupling within the detection circuits can cause problems. A loss of sensitivity or accuracy occurs as the operating frequency is increased.

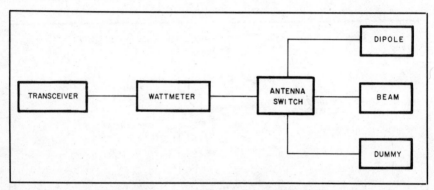

Figure 4-9—A wattmeter is connected at the transmitter output to measure power.

Another type of wattmeter is a **directional wattmeter**. There are two types of directional wattmeters. One type of wattmeter has one meter that reads forward power and another meter to read reflected power. The other has a single meter that can be switched to read either forward or reflected power.

You can use a directional wattmeter to measure the output power from your transmitter. Remember that the reflected power must be subtracted from the forward reading to give you the true forward power. The power reflected from the antenna will again be reflected by the transmitter. This power adds to the forward power reading on the meter. Let's say your transmitter power output is 100 watts and 10 watts are reflected from the antenna. Those 10 watts will be added into the forward reading on your meter when they are reflected from the transmitter. Your wattmeter reading would be incorrect. To find true forward power you must subtract the reflected measurement from the forward power measurement:

True forward power = Forward power reading − Reflected power reading

(Equation 4-1)

For example, suppose you have a wattmeter connected in the line from your transmitter. It gives a forward-power reading of 85 watts and a reflected-power reading of 15 watts. What is the true power out of your transmitter?

True forward power = 85 W − 15 W = 70 watts.

[Now turn to Chapter 10 and study exam questions 3AD-5-1.1 through 3AD-5-1.4, 3AD-5-2.1 and 3AD-5-2.2. Review this section as needed.]

Signal Generators

Marker Generators

The **marker generator** produces a radio frequency signal with a frequency that doesn't change. In its simplest form, the marker generator is a high-stability oscillator. It is usually built into the receiver. This oscillator generates a series of signals that mark the exact edges of the amateur bands (and subbands, in some cases). The marker generator does this by oscillating at a low frequency that has harmonics falling on the desired frequencies. Most marker generators put out harmonics at 25, 50 or 100-kHz intervals. Since marker generators normally use crystal oscillators, they are often called crystal calibrators.

The marker generator is very useful. You can calibrate your receiver with it. You can then determine your transmitter frequency. First, turn on your marker generator (calibrator). This injects the calibrator signal into the receiver circuit. Then set your receiver dial on the proper frequency marks. Calibration instructions are usually provided in your receiver operator's manual. When the receiver dial is calibrated, put your transmitter in the tune or spot position and read the frequency on the receiver dial.

The marker frequencies must be accurate. Now you know that the transmitter frequency is between the markers that show the ends of the band (or subband). In addition, the transmitter frequency must not be too close to the edge of the band (or subband). If it is, the sideband frequencies, especially in a voice transmission, will extend over the edge.

Signal Generators

A **signal generator** produces a stable, low-level signal that can be set to a specific frequency. There are two different types of signal generators. Audio frequency signal generators create signals in the audio range. Radio frequency signal generators provide radio frequency signals. These signals can be used to align or adjust circuits for optimum performance. One common use of a signal generator is in the alignment of receivers. The generator can be adjusted to the desired signal frequency. Then the associated circuits can be adjusted for best operation. This is indicated by the appropriate output meter (maximum signal strength, for instance).

Sometimes, a band of frequencies must be covered to chart the frequency response of a filter. For this application, a swept frequency generator is used. A swept frequency generator automatically sweeps back and forth over a selected range of frequencies.

You can use a signal generator to adjust the filter circuits in your transmitter. When you do, **a dummy load** (or **dummy antenna**, as it is sometimes called) must be connected to the transmitter output. The dummy load acts as a constant load for the transmitter. It replaces the antenna without radiating a signal. More about dummy loads later.

[Now turn to Chapter 10 and study questions 3AD-6.1 through 3AD-6.3, 3AD-7.1 and 3AD-7.2. Review this section as needed.]

ANTENNA MEASUREMENTS

A properly operating antenna system is essential for a top-quality amateur station. If you build your own antennas, you must be able to tune them for maximum operating efficiency. Even if you buy commercial antennas, there are tuning adjustments that must be made because of differences in mounting location. Height above ground and proximity to buildings and trees will have some effect on antenna operating characteristics.

You must be able to measure your antenna's performance. After installation, you should be able to periodically monitor your antenna system for signs of problems, and troubleshoot failures as necessary.

Impedance Match Indicators

A useful metering device is the **reflectometer** (SWR meter). A reflectometer is a device used to measure something called standing-wave ratio, or SWR. SWR is a measure of the relationship between the amount of power traveling to the antenna and the power reflected back to the transmitter. The reflection is caused by impedance mismatches in the antenna system. The energy going from the transmitter to the antenna is represented by the forward, or incident, voltage. The energy reflected by the antenna is represented by the reflected voltage. A bridge circuit separates the incident and reflected voltages for measurement purposes. This is sufficient for determining SWR. Bridges designed to measure SWR are called reflectometers or SWR meters.

A reflectometer is placed in the transmission line between the transmitter and antenna. In the most common amateur application, the reflectometer is connected between the transmitter and Transmatch, as shown in Figure 4-10. A Transmatch is used to cancel out the capacitive or inductive component in the antenna impedance. With a Transmatch you can match the impedance of your antenna system to the transmitter output, normally 50 ohms. The reflectometer is used to indicate minimum reflected power as the Transmatch is adjusted. This indicates when the antenna system (including feed line) is matched to the transmitter output impedance.

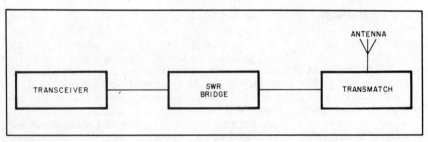

Figure 4-10—A reflectometer or SWR bridge should be connected between the transmitter and Transmatch in order to adjust the Transmatch properly.

To measure the impedance match between an antenna and the feed line, place the SWR meter at the antenna feed point. If you put the meter at the transmitter end, feed line losses make SWR readings look lower than they really are. More information on antennas and feed lines will be covered in Chapter 9.

[Now study questions 3AD-8-1.1, 3AD-8-1.2, 3AD-8-2.1, and 3AD-8-2.2 in Chapter 10. Review any material necessary.]

STATION ACCESSORIES

There are many other station accessories available. They make the difference between simply operating a radio and having a full awareness of the communications medium called Amateur Radio. In this section we will look at some of these accessories. You may find them useful when you upgrade to a Technician class license.

Dummy Loads or Dummy Antennas

A **dummy load** (sometimes called a **dummy antenna**) is a resistor. It has impedance characteristics that allow it to take the place of a regular antenna system for test purposes. You can use a switch to connect the antenna or dummy load to the transmitter. See Figure 4-11. With a dummy load connected to the transmitter, you can make tests without having a signal go out over the air. A dummy load is especially handy for transmitter testing. FCC rules strictly limit the amount of on-the-air testing that may be done (see Section 97.93 of the amateur rules).

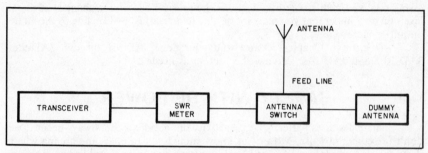

Figure 4-11—A dummy load can be switched in to provide a nonradiating load for transmitter adjustments.

For transmitter tests, remember that a dummy load is a resistor. It must be capable of safely dissipating the entire power output of the transmitter. A dummy load should provide the transmitter with a perfect load. This usually means that it has a pure resistance (with no reactance) of approximately 50 ohms. The resistors used to construct dummy loads must be noninductive. Composition resistors are usable, but wire-wound resistors are not. A single high-power resistor is best. Several lower-power resistors can be connected in parallel to obtain a 50-ohm load capable of dissipating high power.

The S Meter

S meters are used on receivers to indicate the strength of a received signal. Signal-strength meters are useful when there is a need to make comparative readings.

S meters are calibrated in S units from S1 to S9; above S9 they are calibrated in decibels (dB) as shown in Figure 4-12. In the 1940's at least one manufacturer made an attempt to establish some significant numbers for S meters. S9 was to be equal to a signal level of 50 microvolts, with each S unit equal to 6 dB. This means that for an increase of one S unit, the received signal power would have to increase by

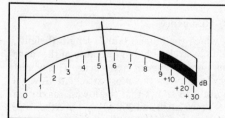

Figure 4-12—An S-meter is calibrated in S-units up to S-9. Above S-9 it is calibrated in decibels (dB) above S-9.

a factor of four. Such a scale is useful for a theoretical discussion. Real S-meter circuits fall far short of this ideal for a variety of reasons.

Suppose you were working someone who was using a 25-watt transmitter, and your S meter was reading S7. If the other operator increased power to 100 watts (an increase of four times), your S meter would read S8. If the signal was S9, the operator would have to multiply his power 10-fold to make your S meter read 10 dB over S9.

S meters on modern receivers may or may not respond in this manner. S-meter operation is based on the output of the automatic gain-control (AGC) circuitry; the S meter measures the AGC voltage. As a result, every receiver S meter responds differently; no two S meters will give the same reading. The S meter is useful for giving relative signal-strength indications, however. You can see changes in signal levels on an S meter that you may not be able to detect just by listening to the audio output level.

[Now turn to Chapter 10 and study questions 3AD-9.1 through 3AD-9.6, 3AD-10.1 and 3AD-10.2. Review this section as needed.]

SAFETY WITH RF POWER

Amateur Radio is basically a safe activity but accidents can always occur if we don't use common sense. Most of us know enough not to place an antenna where it can fall on a power line. We don't insert our hand into an energized linear amplifier, or climb a tower on a windy day. We also should not venture overexposure to RF energy. Large amounts of RF energy *can* be harmful to people because it heats body tissues. The effects depend on the wavelength, energy density of the incident RF field, and on other factors such as polarization.

The most susceptible parts of the body are the tissues of the eyes. They don't have heat-sensitive receptors to warn us of the danger before the damage occurs. Symptoms of overexposure may not appear until after irreversible damage has been done. Though the problem should be taken seriously, with reasonable precaution, Amateur Radio operation can be safe.

Safe Exposure Levels

In recent years scientists have devoted a great deal of effort to determining safe RF-exposure limits. The problem is very complex. It's not surprising that some changes in the recommended levels have occurred as more information has become available. The American Radio Relay League believes that the latest "Radio Frequency Protection Guide of the American National Standards Institute (ANSI)" is the best available protection standard; it took nearly five years to formulate and has undergone repeated critical review by the scientific community. This 1982 guide recognizes the phenomenon of whole-body or geometric resonance and establishes a frequency-dependent maximum permissible RF exposure level. Exposure levels are expressed in terms of power density. Power density is measured in milliwatts per square centimeter. It is a measure of the radio frequency power that strikes a person per square centimeter of body surface.

Whole-Body Resonance

Resonance occurs at frequencies for which the human body's length (height), if parallel to the antenna (vertical), is about 0.4 wavelength long. Because of the range of human heights, the resonant region spans a broad range of frequencies. This whole-body resonance establishes the frequency range for the most stringent (lowest) maximum permissible exposure level. Figure 4-13 shows that the lowest maximum exposure level is 1 mW/cm^2 for frequencies between 30 and 300 MHz. On either side of those "corner" frequencies the rise is gradual. At 3 MHz the maximum permissible exposure level is 100 mW/cm²; at 1500 MHz and above, 5 mW/cm². The valley region includes some active amateur bands (10, 6 and 2 meters) as well as all FM and TV broadcasting. The rationale for specifying a constant 5 mW/cm² above 1500 MHz takes into consideration that the extremely short wavelengths don't penetrate very deep into body tissue.

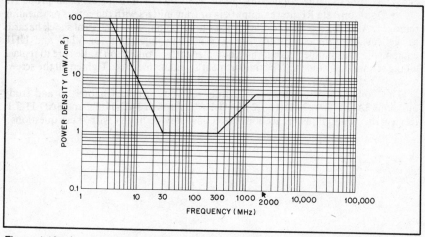

Figure 4-13—American National Standards Institute Radio Frequency Protection Guide for whole-body exposure of human beings.

RF Safety Guidelines

Take the time to study and follow these general guidelines to minimize your exposure to RF fields. Most of these guidelines are just common sense and good amateur practice.

• In high-power operation in the HF and VHF region, keep the antenna away from people. Humans should not be allowed within 10 to 15 feet of vertical antennas. This is especially important with higher power, high-duty-cycle operation (such as FM or RTTY). Amateur antennas that are mounted on towers and masts, away from people, pose no exposure problem.

• When using mobile equipment with 10-W RF power output or more, do not transmit if anyone is standing within 2 feet of the antenna.

• When using hand-held transceivers with RF power output of several watts or more, maintain at least 1 to 2 inches separation between the antenna and your forehead. ANSI Standards recommend power outputs of no more than seven watts for hand-held radios. This is because the antenna is so close to the operator's head.

• Never touch an antenna that has RF power applied. Be sure RF power is off—and stays off—before working on or adjusting an antenna. Also, make sure any nearby antennas are deactivated. Never have someone else transmit into the antenna and monitor the SWR while you are making adjustments. When matching an antenna, the correct procedure is to turn the transmitter off, make the adjustment, and then back away to a safe distance before turning the transmitter on again to check your work.

• During transmissions, never point a high-gain UHF antenna (such as a parabolic dish) toward people or animals.

• Never look into the open end of a UHF waveguide feed line that is carrying RF power. Never point the open end of a UHF waveguide that is carrying RF power toward people or animals. Make sure that all waveguide connections are tightly secured.

• Confine RF radiation to the antenna, where it belongs. Provide a good earth ground for your equipment. Poor feed line and improperly installed connectors can be a source of unwanted radiation. Use only good-quality, well-constructed coaxial cable. Be sure that the connectors are of good quality and install them properly.

• Don't operate RF power amplifiers or transmitters with the covers or shielding removed. Always keep all shielding in place to avoid potential electrical shock hazards as well as RF safety hazards. This is especially important for VHF and UHF equipment. Reassemble transmitting equipment after working on it. Be sure to replace all the screws that hold the RF compartment shielding in place. Tighten all the screws securely before applying power to the equipment.

[This completes your study of Chapter 4. Now turn to Chapter 10 and study questions 3AD-11-1.1, 3AD-11-1.2 , 3AD-11-2.1 through 3AD-11-2.5 and 3AD-11-3.1. Review the material in this section if you have any difficulty with these questions.]

Key Words

Alternating current (ac)—Electrical current that flows first in one direction in a wire and then in the other direction. The applied voltage changes polarity and causes the current to change direction. This direction reversal continues at a rate that depends on the frequency of the ac.

Capacitor—An electrical component composed of two or more conductive plates separated by an insulating material. A capacitor stores energy in an electrostatic field.

Coil—A conductor wound into a series of loops.

Core—The material in the center of a coil. The material used for the core affects the inductance value of the coil.

Current—A flow of electrons in an electrical circuit.

Dielectric—The insulating material used between plates in a capacitor.

Direct current (dc)—Electrical current that flows in one direction only.

Electromotive force (EMF)—The force or pressure that pushes a current through a circuit.

Farad—The basic unit of capacitance.

Henry—The basic unit of inductance.

Induced EMF—A voltage produced by a change in magnetic lines of force around a conductor. When a magnetic field is formed by current in the conductor, the induced voltage always opposes changes in that current.

Inductor—An electrical component usually composed of a coil of wire wound on a central core. An inductor stores energy in a magnetic field.

Ohm—The basic unit of resistance.

Ohm's Law—A basic law of electronics, it gives a relationship between voltage, resistance and current ($E = IR$).

Parallel circuit—An electrical circuit in which the electrons follow more than one path in going from the negative supply terminal to the positive terminal.

Resistance—The ability to oppose an electrical current.

Series circuit—An electrical circuit in which all the electrons must flow through every part of the circuit. There is only one path for the electrons to follow.

Voltage—The force or pressure (EMF) that pushes a current through a circuit.

Chapter 5

Electrical Principles

T o pass your FCC Element 2 written test and receive a Novice license, you had to learn some very basic radio theory. For your Element 3A test, you will need to understand a few more electrical principles. This chapter will introduce you to the basic concepts of resistance, capacitance and inductance.

Ohm's Law is an important tool for analyzing circuits. We'll look at some applications of this basic electronics principle. We'll look at methods for finding the total value of parallel and series combinations of resistors, capacitors and inductors.

Be sure to turn to Chapter 10 and study the appropriate FCC questions when the text directs you to do so. This will show you if you are progressing smoothly, or if you need to do a little extra studying. We cannot tell you everything about a particular topic in one chapter. You may want to use some other reference books for additional information. *The ARRL Handbook for the Radio Amateur* is a good place to start.

E = VOLTAGE

Electrons need a push to get them moving. We call the force that pushes electrons though a circuit **electromotive force (EMF)**. EMF is similar to water pressure in a pipe. See Figure 5-1. More pressure causes more water to flow through the pipe. Greater EMF causes a greater flow of electrons in a circuit. See Figure 5-2.

EMF is measured in volts, so it is usually called **voltage**. A voltage-measuring instrument is a voltmeter. More voltage applied to a circuit makes more electrons flow through the circuit.

I = CURRENT

The word **current** comes to us from a Latin word meaning "to run" and it implies movement. When someone speaks of the current in a river, we think of a flow of water. A current of air can be a light breeze or a hurricane. A current of electricity is a flow of electrons.

Current is indicated in equations and diagrams by the letter I (from the French word *intensité). We measure current in amperes, and the measuring device is an ammeter. An ampere is often an inconveniently large unit. We can measure current in milliamperes (one thousandth of an ampere) or microamperes (one millionth of an ampere). The abbreviation for amperes (sometimes referred to informally as amps) is A; milliamperes mA and microamperes μA.*

When considering current, it is natural to think of a single, constant force causing the electrons to move. When this is so, the electrons always move in the same

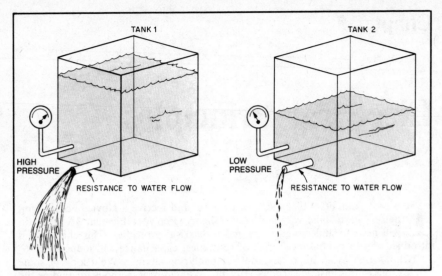

Figure 5-1—Since the water-tank outlet pipes are identical in size, they present an identical amount of resistance to water flow. The pressure at the outlets is different because of the difference in water level. This is similar to two separate electrical circuits with identical amounts of resistance and different supply voltages. The higher pressure of Tank 1 causes more water to flow over a given time, just as a higher voltage causes larger electrical currents.

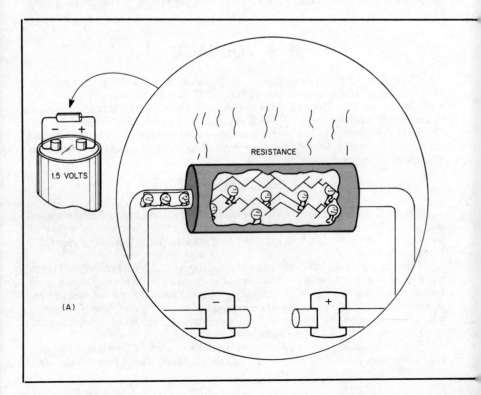

(A)

direction through a circuit. A circuit is made up of conductors connected in a continuous loop. A constant current flowing in a circuit is **direct current**, abbreviated **dc**. This is the type of current furnished by batteries and by certain types of generators.

R = RESISTANCE

Resistance is something that opposes motion. All materials have electrical resistance. There is always some opposition to the flow of electrons through the material. The resistance of different materials varies widely. Some materials have very high resistance; we call these materials good insulators, or poor conductors. Other materials have very low resistance; we call these materials poor insulators, or good conductors. The basic unit of resistance is the **ohm**, and a resistance-measuring instrument is an ohmmeter. The symbol for an ohm is Ω, the Greek letter capital omega.

We can make devices with a range of opposition to the flow of electrons in a circuit. They are between the extremes of a very good insulator and a very good conductor. We call such devices resistors. We will use the analogy of water flowing

---

Figure 5-2—At A, 1.5 volts supplies electrons in the wire and resistive material with energy to try to "push" their way through the circuit from one battery terminal to the other. The number of electrons that can move in a given amount of time is limited by resistance. The electrons find it easy going in the low-resistance wire, but have to expend more energy in passing through the higher resistance. This energy is released as heat. The 12-volt battery at B has the electrical force needed to move many more electrons. Much more heat is given off by the resistive material.

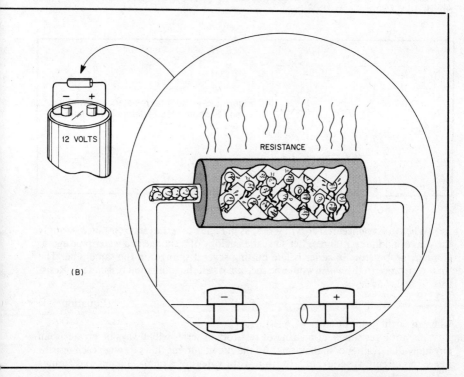

in a pipe to help us understand electron flow. These materials act like a pipe with a sponge in it. Some water will still flow in the pipe, but the flow will be reduced. The primary function of a resistor is to limit electron flow (current). Greater resistance causes greater reduction in current. If we want to control the current reduction over a certain range, we can use a variable resistor. A variable resistor can be made to change resistance over a range. We can then control the flow of electrons from a maximum to a minimum over this range.

This current reduction has a price, however. The energy lost by the electrons as they flow through a resistor must be dissipated in some way. The energy converts into heat, and the resistor gets warm. If the current is higher than the resistor can handle, the resistor will get very hot and may even burn.

There is another kind of opposition to electrical current called reactance. Later in this book you will learn about capacitors and inductors. Both of these components can oppose electrical currents without generating heat. This property is called reactance. A perfect resistance does not include reactance in its opposition to electric current.

[Turn to Chapter 10 and study exam questions 3AE-1-1.1, 3AE-1-2.1 and 3AE-1-2.2. Review this section as needed.]

Resistors in Series and Parallel

Sometimes it's necessary to calculate the total resistance of resistors connected in a **series circuit**. Resistors in series are connected end to end like a string of sausages. At other times you must calculate total resistance in a **parallel circuit**. In a parallel circuit, resistors are side by side like a picket fence. There will be times when you need a certain amount of resistance somewhere in a circuit. There may be no standard resistor value that will give the necessary resistance. Sometimes you may not have a certain value on hand. By combining resistors in parallel or series you can obtain the desired value.

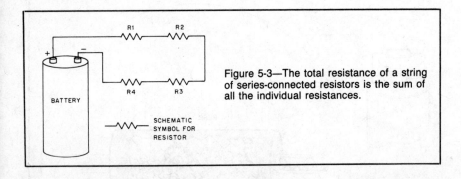

Figure 5-3—The total resistance of a string of series-connected resistors is the sum of all the individual resistances.

When you connect resistors in series, as in Figure 5-3, the total resistance is simply the sum of all the resistances. Let's use the analogy of a sponge in a water pipe again. Connecting resistors in series is like putting several sponges in the same pipe. The total resistance to the water would be the sum of all the individual resistances. Resistors in series add.

$$R_{TOTAL} = R_1 + R_2 + R_3 + \ldots + R_n \qquad \text{(Equation 5-1)}$$

where n is the total number of resistors.

The total resistance of a string of resistors in series will always be greater than any individual resistance in the string. All the circuit current flows through each resistor in a series circuit. A series circuit with resistor values of 2 ohms, 3 ohms and 5 ohms

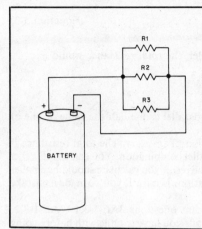

Figure 5-4—Parallel-connected resistors. In this circuit, the electrical current will split into three separate paths through the resistors. Each resistor will be exposed to the full battery voltage. Current through each individual resistor is independent of the other resistors. For example, if R2 and R3 are changed to a value different than the original value, the current in R1 would remain unchanged. Even if R2 and R3 are removed, the current in R1 remains unchanged.

would have a total resistance of 10 ohms.

In a parallel circuit, things are a bit different. When we connect two or more resistors in parallel, more than one path for current exists in the circuit. See Figure 5-4. This is like connecting another water pipe into our water-pipe circuit. When there is more than one path, more water can flow during a given time. With more than one resistor, more electrons can flow. This means there is a greater current.

The formula for calculating the total resistance of resistors connected in parallel is:

$$R_{TOTAL} = \cfrac{1}{\cfrac{1}{R_1} + \cfrac{1}{R_2} + \cfrac{1}{R_3} + \ldots + \cfrac{1}{R_n}}$$

(Equation 5-2)

where n is the total number of resistors. For example, if we connect three 100-ohm resistors in parallel, their total resistance is:

$$R_{TOTAL} = \cfrac{1}{\cfrac{1}{100} + \cfrac{1}{100} + \cfrac{1}{100}}$$

$$R_{TOTAL} = \cfrac{1}{\cfrac{3}{100}} = \frac{1}{0.03} = 33.33 \text{ ohms}$$

If we connect a 25-ohm resistor, a 100-ohm resistor and a 10-ohm resistor in parallel, our equation becomes:

$$R_{TOTAL} = \cfrac{1}{\cfrac{1}{25} + \cfrac{1}{100} + \cfrac{1}{10}}$$

$$R_{TOTAL} = \frac{1}{0.04 + 0.01 + 0.1} = \frac{1}{0.15} = 6.67 \text{ ohms}$$

To calculate the total resistance of two resistors in parallel, Equation 5-2 reduces to the "product over sum" formula. Divide the product of the two resistances by their sum:

$$R_{TOTAL} = \frac{R_1 \times R_2}{R_1 + R_2} \qquad \text{(Equation 5-3)}$$

If we connect two 50-ohm resistors in parallel, the total resistance would be:

$$R_{TOTAL} = \frac{50 \times 50}{50 + 50} = \frac{2500}{100} = 25 \text{ ohms}$$

The total resistance of two equal resistors in parallel is one-half the value of one of the resistors.

Now let's look at any parallel combination of resistors. The total resistance is always less than the smallest value of the parallel combination. You can use this fact to make a quick check of your calculations. The result you calculate should be smaller than the smallest value in the parallel combination. If it isn't, you've made a mistake somewhere!

[Now turn to Chapter 10 and study exam questions 3AE-1-3.1, 3AE-1-3.2, 3AE-1-4.1 and 3AE-1-4.2. Review any equations you have trouble with before going on.]

OHM'S LAW

Resistance, voltage and current are related in a very important way by what is known as **Ohm's Law**. Georg Simon Ohm was a German scientist. He discovered that the voltage drop across a resistor is equal to the resistance multiplied by the current through the resistor. In symbolic terms, Ohm's Law can be written:

$$E = I \times R \qquad \text{(Equation 5-4)}$$

where E is voltage, I is current and R is resistance.

By simply rearranging the terms, we can show that the current is equal to the voltage drop divided by the resistance:

$$I = \frac{E}{R} \qquad \text{(Equation 5-5)}$$

and that the resistance equals the voltage drop divided by the current:

$$R = \frac{E}{I} \qquad \text{(Equation 5-6)}$$

These relationships are true when quantities are expressed as volts, ohms, and amperes. For example, given milliamperes and volts, convert milliamperes to amperes before calculating to obtain the result expressed in ohms. (The calculation can be made directly without converting to amperes to obtain an answer expressed in kilohms. However, determining what kind of units should be used in expressing the answer is a potential source of error.)

Ohm's Law is an important mathematical relationship. You should memorize and be able to use it. A simple way to remember the Ohm's Law equation is shown in Figure 5-5. Draw the "Ohm's Law circle" shown. You can find the equation for any of the three quantities in the relationship.

To use the circle, just cover the letter you want to solve the equation for. If the two remaining letters are across from each other, you must multiply them. If the two letters are one on top of the other, you must divide the top letter by the bottom.

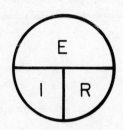

Figure 5-5—The "Ohm's Law Circle." Cover the letter representing the unknown quantity to find a formula to calculate that quantity. For example, if you cover the I, you are left with E / R.

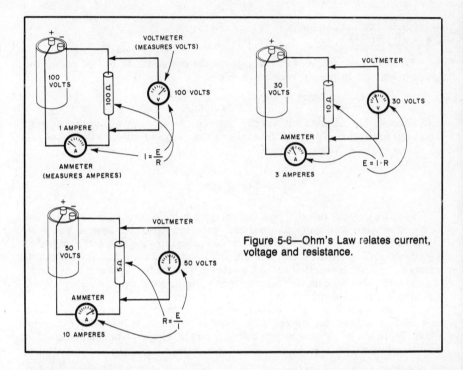

Figure 5-6—Ohm's Law relates current, voltage and resistance.

For example, if you cover I, you have "E over R" (Equation 5-5). Cover E, and you have "I times R" (Equation 5-4).

When you know any two quantities in Ohm's Law, you can always find the third. See Figure 5-6. Suppose you have a circuit where the resistance is 10 ohms and the current is 3 amperes. To find the voltage, multiply the resistance times the current.

$E = I \times R$

$E = 3$ amperes $\times$ 10 ohms

$E = 30$ volts

In a circuit with 100 volts applied to a 100-ohm resistor, you would use Equation

5-5 to find the current. Substituting the values for our problem into the formula, we have:

$$I = \frac{E}{R}$$

$$I = \frac{100 \text{ volts}}{100 \text{ ohms}} = 1 \text{ ampere}$$

In a circuit with 50 volts applied and 10 amperes of current flowing, we use Equation 5-6 to find the resistance:

$$R = \frac{E}{I}$$

$$R = \frac{50 \text{ volts}}{10 \text{ amperes}} = 5 \text{ ohms}$$

To check our work we can always work the problem back to find one of the given quantities. In the last problem, using $E = I \times R$:

$$E = 10 \text{ amperes} \times 5 \text{ ohms} = 50 \text{ volts}$$

and we prove that our math was correct! You won't go wrong by writing the equation you want to use first. Then, substitute the values given and solve for the unknown, as shown in the previous examples.

Power

Suppose we want to know how fast energy is being consumed in a circuit. Can you operate your Amateur Radio station, lights, fan, air conditioner or heater on the same house circuit? Do you know what size generator your club needs for portable operations? To answer these questions you will need to know how fast the equipment and appliances use electrical energy. Power is defined as the time rate of energy consumption. The basic unit for measuring power is the watt, named for James Watt, the inventor of the steam engine.

Current is a measure of how many electrons are moving in an electrical circuit in a given period of time. Voltage is a measure of the force needed to move the electrons. There is a very simple way to calculate power using current and voltage. We use the equation:

$$P = I \times E \quad\quad\quad\quad\quad\quad\quad\quad\quad\quad\quad\quad\quad\quad \text{(Equation 5-7)}$$

where
     P = power, measured in watts
     E = the EMF in volts
     I = the current in amperes

To calculate the power in a circuit, multiply volts times amperes. For example, in Figure 5-7 a 12-V battery is providing 3 A of current to a light bulb that is operating normally. Using Equation 5-7, we can find the power rating for the bulb by multiplying the voltage by the current:

$$P = I \times E = 3 \text{ A} \times 12 \text{ V} = 36 \text{ watts}$$

If the local radio club needs to supply 20 amperes at 120 volts from a portable generator, then the generator must be able to product at least 2400 watts (2.4 kilowatts).

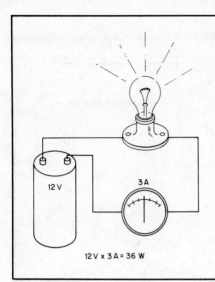

Figure 5-7—Applied voltage multiplied by current equals power.

12 V

3 A

12 V x 3 A = 36 W

$P = I \times E = 20\ A \times 120\ V = 2400$ watts

What if we want to know the power dissipated in a 1000-ohm resistor that has 0.3 ampere flowing through it? To use Equation 5-7 we need to know the applied voltage—but we can find that with Ohm's Law:

$E = I \times R = 0.3\ A \times 1000\ \Omega = 300$ volts

Now we can use Equation 5-7:

$P = I \times E = 300\ V \times 0.3\ A = 90$ watts

[Now turn to Chapter 10 and study questions 3AE-2.1 through 3AE-2.9. You should have a thorough understanding of Ohm's Law before you go on to the next section. Review any material you have trouble with before proceeding.]

INDUCTANCE

The motion of electrons produces magnetism. Every electric current creates a magnetic field around the wire in which it flows. Like an invisible tube, the magnetic field is positioned in concentric circles around the conductor. See Figure 5-8A. The field is established when the current flows, and collapses back into the conductor when the current stops. The field increases in strength when the current increases and decreases in strength as the current decreases. The force produced around a straight piece of wire by this magnetic field is usually negligible. When the same wire is formed into a **coil**, the force is much greater. In coils, the magnetic field around each turn also affects the other turns. Together, the combined forces produce one large magnetic field, as shown in Figure 5-8B. Much of the energy in the magnetic field concentrates in the material in the center of the coil (the **core**). Most practical **inductors** consist of a length of wire wound on a core.

An inductor stores energy in a magnetic field. Magnetic fields can also set electrons in motion. When a magnetic field increases in strength, the voltage of a conductor within that field increases. When the field strength decreases, so does the voltage.

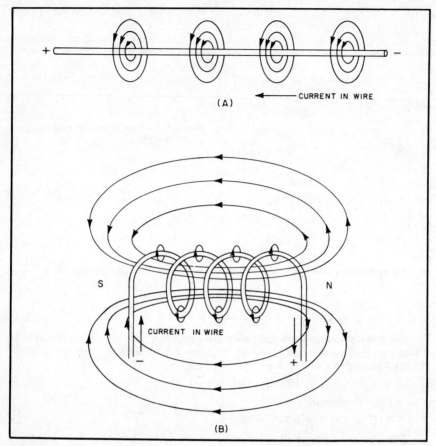

Figure 5-8—A magnetic field surrounds a wire with a current in it. If the wire is formed into a coil, the magnetic field becomes much stronger as the lines of force reinforce each other.

Let's apply a dc voltage to an inductor. As there is no current to start with, the current begins to increase when the voltage is applied. This will establish an electrical current in the inductor. The voltage induced by the magnetic field of the inductor opposes the applied voltage. Therefore, the inductor will oppose the increase in current. This is a basic property of inductors. Any changes in current through the inductor, whether increasing or decreasing, are opposed. A voltage is induced in the coil that opposes the applied voltage, and tries to prevent the current from changing. This is called **induced EMF** (voltage) or back EMF. Gradually a current will be produced by the applied voltage. (The term "gradually" is relative. In radio circuits, the time needed to produce the current in the circuit is often measured in microseconds.)

The "final" current that flows through the inductor is limited only by any resistance that might be in the circuit. There is very little resistance in the wire of most coils. The current will be quite large if there is no other resistance.

In the process of getting this current to flow, energy is stored. This energy is in the form of a magnetic field around the coil. When the applied current is shut off, the magnetic field collapses. The collapsing field returns energy to the circuit as a momentary current that continues to flow in the same direction as the original current. This current can be quite large for large values of inductance. The induced voltage can rise to many times the applied voltage. This can cause a spark to jump across switch or relay contacts when the circuit is broken to turn off the current. This effect is called "inductive kickback."

When an **ac** voltage is applied to an inductor, the current through the inductor will reverse direction every half cycle. This means the current will be constantly changing. The inductor will oppose this change. Energy is stored in the magnetic field while the current is increasing during the first half cycle. This energy will be returned to the circuit as the current starts to decrease. A new magnetic field will be produced during the second half cycle. The north and south poles of the field will be the reverse of the first half cycle. The energy stored in that field will be returned to the circuit as the current again starts to decrease. Then a new magnetic field will be produced on the next half cycle. This process keeps repeating, as long as the ac voltage is applied to the inductor.

Factors That Determine Inductance

The inductance of a coil determines several circuit conditions. One is the amount of opposition to changes in current. Another is the amount of energy stored in the magnetic field. The back EMF induced in the coil also depends on the inductance. In turn, the inductance of a coil, usually represented by the symbol L, depends on four things:

1) the type of material used for the core (permeability of the material), and its size and location in the coil;

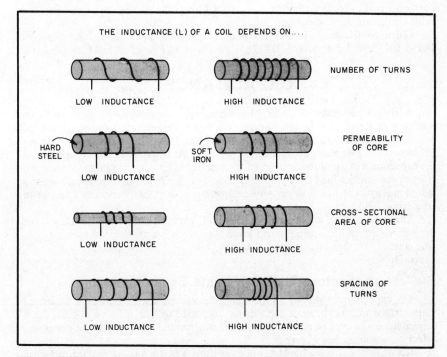

Figure 5-9—The value of an inductor depends on the material used in the core, the number of windings, and the length and diameter of the coil.

2) the number of turns used to wind the coil;
3) the length of the coil (spacing of turns);
4) the diameter of the coil (cross-sectional area)
Changing any of these factors changes the inductance. See Figure 5-9.

The basic unit of inductance is the **henry**, named for the American physicist Joseph Henry. The henry is often too large for practical use in measurements. We use the millihenry (10^{-3}) abbreviated mH, or microhenry (10^{-6}) abbreviated μH.

Inductors in Series and Parallel

In circuits, inductors combine like resistors. The total inductance of several inductors connected in series is the sum of all the inductances:

$$L_{TOTAL} = L_1 + L_2 + L_3 + \ldots + L_n \qquad \text{(Equation 5-8)}$$

where n is the total number of inductors.

For parallel-connected inductors:

$$L_{TOTAL} = \frac{1}{\dfrac{1}{L_1} + \dfrac{1}{L_2} + \dfrac{1}{L_3} + \ldots + \dfrac{1}{L_n}} \qquad \text{(Equation 5-9)}$$

The total inductance of several inductors connected in parallel is less than the smallest inductance value in the combination. You should recognize this equation by now, and realize that for two parallel-connected inductors, Equation 5-9 reduces to:

$$L_{TOTAL} = \frac{L_1 \times L_2}{L_1 + L_2} \qquad \text{(Equation 5-10)}$$

For two equal inductors connected in parallel, the total value will be one-half the value of one of the components.

[Turn to Chapter 10 and study questions 3AE-3-1.1, 3AE-3-2.1 through 3AE-3-2.4, 3AE-3-3.1, 3AE-3-3.2, 3AE-3-4.1 and 3AE-3-4.2. Review as needed.]

CAPACITANCE

A simple **capacitor** is formed by separating two conductive plates with an insulating material, or **dielectric**. Connect one plate to the positive terminal of a voltage source. Connect the other plate to the negative terminal. We can build up a surplus of electrons on one plate, as shown in Figure 5-10. At some point, the voltage across the capacitor will equal the applied voltage, and the capacitor is said to be charged. If we then connect a load to the capacitor, it will discharge through the load, releasing stored energy. The basic property of a capacitor is this ability to store a charge in an electric field.

The basic unit of capacitance is the **farad**, named for Michael Faraday. Like the henry, the farad is usually too large a unit for practical measurements. For convenience, we use microfarads (10^{-6}), abbreviated μF, or picofarads (10^{-12}), abbreviated pF.

Factors That Determine Capacitance

The capacitance value of a capacitor is determined by three factors. Increased plate-surface area will increase capacitance. Increased spacing between plates reduces capacitance. The type of insulating material (dielectric) used between the plates will affect capacitance. See Figure 5-11.

Reducing the spacing between plates or using a better insulator as the dielectric will increase the capacitance. For a given plate size, using multiple plates in the

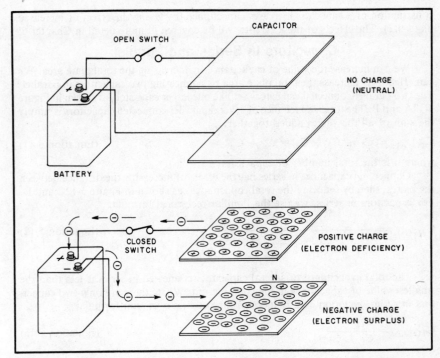

Figure 5-10—When a voltage is applied to a capacitor, an electron surplus (negative charge) builds up on one plate, while an electron deficiency forms on the other plate to produce a positive charge.

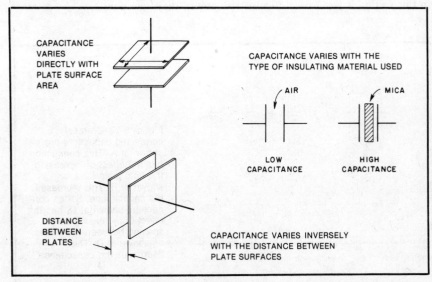

Figure 5-11—The capacitance of a capacitor depends on the area of the plates, the distance between the plates and the type of dielectric material used.

construction of a capacitor increases capacitance (by giving the effect of increased plate size). The effects of these factors will be covered in more detail in Chapter 6.

Capacitors in Series and Parallel

We can increase the value of capacitance by increasing the total plate area. We can effectively increase the total plate area by connecting two capacitors in parallel. The two parallel-connected capacitors act like one larger capacitor, as shown in Figure 5-12A and B. The total capacitance of several parallel-connected capacitors is simply the sum of all the values added together:

$$C_{TOTAL} = C_1 + C_2 + C_3 + ... + C_n \qquad \text{(Equation 5-11)}$$

where n is the total number of capacitors.

Connecting capacitors in series has the effect of increasing the distance between the plates, thereby reducing the total capacitance, as shown in Figure 5-12C and D. For capacitors in series, we use the familiar reciprocal formula:

$$C_{TOTAL} = \cfrac{1}{\cfrac{1}{C_1} + \cfrac{1}{C_2} + \cfrac{1}{C_3} + ... + \cfrac{1}{C_n}} \qquad \text{(Equation 5-12)}$$

The total capacitance of several capacitors connected in series is less than the smallest value of capacitance in the combination. Where there are only two capacitors in series, equation 5-12 reduces to the product over sum formula:

$$C_{TOTAL} = \frac{C_1 \times C_2}{C_1 + C_2} \qquad \text{(Equation 5-13)}$$

As we discovered using this formula for two equal resistors in parallel, two equal capacitors connected in series total half the value of either single capacitor.

[This completes your study of Chapter 5. Now turn to Chapter 10 and study exam questions 3AE-4-1.1, 3AE-4-2.1 through 3AE-4-2.4, 3AE-4-3.1, 3AE-4-3.2, 3AE-4-4.1 and 3AE-4-4.2. Before proceeding to the next chapter, review the material in this section if you have difficulty with any of these questions.]

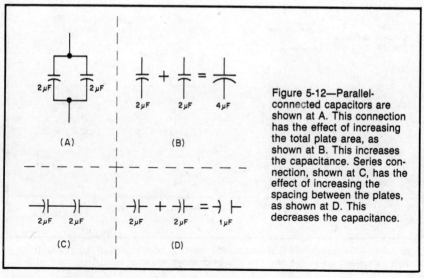

Figure 5-12—Parallel-connected capacitors are shown at A. This connection has the effect of increasing the total plate area, as shown at B. This increases the capacitance. Series connection, shown at C, has the effect of increasing the spacing between the plates, as shown at D. This decreases the capacitance.

Key Words

Breakdown voltage—The voltage at which an insulating material will conduct current.

Capacitor—An electronic component composed of two or more conductive plates separated by an insulating material. A capacitor stores energy in an electrostatic field.

Carbon-composition resistor—An electronic component designed to limit current in a circuit; made from ground carbon mixed with clay.

Carbon-film resistor—A resistor made by putting a gaseous carbon deposit on a round ceramic form.

Ceramic capacitor—An electronic component composed of two or more conductive plates separated by a ceramic insulating material.

Color code—A system where numerical values are assigned to various colors. Colored stripes are painted on the body of resistors and sometimes other components to show their value.

Core—The material used in the center of a coil. The material used for the core affects the inductance value of the coil.

Dielectric—The insulating material used between the plates in a capacitor.

Dielectric constant—A number used to indicate the relative "merit" of an insulating material. Air is given a value of 1, and all other materials are related to air.

Electric field—An invisible force of nature. An electric field exists in a region of space if an electrically charged object placed in the region is subjected to an electrical force.

Electrolytic capacitor—A polarized capacitor formed by using thin foil electrodes and chemical-soaked paper.

Fixed resistor—A resistor with a fixed nonadjustable value of resistance.

Metal-film resistor—A resistor formed by depositing a thin layer of resistive-metal alloy on a cylindrical ceramic form.

Mica capacitor—A capacitor formed by alternating layers of metal foil with thin sheets of insulating mica.

Mutual coupling—When coils display mutual coupling, a current flowing in one coil will induce a voltage in the other. The magnetic flux of one coil passes through the windings of the other.

Paper capacitor—A capacitor formed by sandwiching paper between thin foil plates, and rolling the entire unit into a cylinder.

Plastic-film capacitor—A capacitor formed by sandwiching thin sheets of Mylar™ or polystyrene between thin foil plates, and rolling the entire unit into a cylinder.

Potentiometer—A resistor whose resistance can be varied continuously over a range of values.

Reactance—The property of an inductor or capacitor (measured in ohms) that impedes current in an ac circuit without converting power to heat.

Resistor—Any material that opposes a current in an electrical circuit. An electronic component specifically designed to oppose current.

Rotor—The movable plates in a variable capacitor.

Stator—The stationary plates in a variable capacitor.

Toroidal inductor—A coil wound on a donut-shaped ferrite or powdered-iron form.

Variable capacitor—A capacitor that can have its value changed within a certain range.

Variable resistor—A resistor whose value can be adjusted over a certain range.

Wire-wound resistor—A resistor made by winding a length of wire on an insulating form.

Chapter 6

Circuit Components

Y ou can understand the operation of most complex electronic circuits. First, you must know some basic information about the parts that make up those circuits. This chapter presents the information about circuit components that you will need to know to pass your Technician class written exam. You will find descriptions of resistors, capacitors and inductors. You can combine these components with other devices to build practical electronic circuits. We describe some of these circuits in Chapter 7.

Study the characteristics of the components described in this chapter. Also, study and be able to identify the schematic symbols for the components. Schematic symbols are used in schematic diagrams. A schematic diagram shows component connections without regard to physical position. Then, turn to the Element 3A questions in Chapter 10 when directed to do so. That will show you where you need to do some extra studying. You should thoroughly understand how these components work. When you do, you can connect them to make a circuit perform a specific task.

RESISTORS

We have seen that current in a circuit is the flow of electrons from one point to another. A perfect insulator would allow no electron flow (zero current), while a perfect conductor would allow infinite electron flow (infinite current). In practice, however, there is no such thing as a perfect conductor or a perfect insulator. Partial opposition to electron flow occurs when the electrons collide with other electrons or atoms in the conductor. The result is a reduction in current, and the conductor produces heat.

Resistors allow us to control the current in a circuit by controlling the opposition to electron flow. As they oppose the flow of electrons they dissipate electrical energy in the form of heat. The more energy a resistor dissipates, the hotter it will become. Figure 6-1A shows the schematic symbol for all nonadjustable resistors. Part B shows a common resistor. You will learn how to interpret the colored stripes on a resistor later in this chapter.

All conductors exhibit some resistance. Table 6-1 shows the resistance in ohms per 1000 feet of some common copper and nickel American wire gauge (AWG) wire sizes. We can see that the smaller the conductor, the greater the resistance. This is just what we would expect. Think of electron flow in a conductor as being like water flow in a pipe. If we reduce the size of the pipe, not as much water can flow through it.

One simple way of producing a resistor is to use a length of wire. For example, we can see from Table 6-1 that the resistance of 1000 feet of number 28 nickel wire is 337 ohms. Therefore, if we need a resistor of 3.4 ohms, we can use 10 feet of this

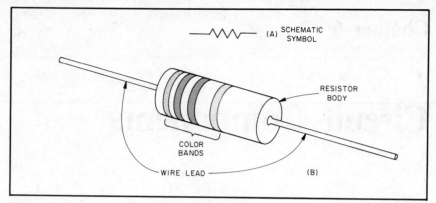

Figure 6-1—Part A shows the schematic symbol for a resistor. Part B shows a typical resistor, including the colored stripes that identify the resistor value. Resistors made in this way range in power rating from 1/8 watt to 2 watts.

Table 6-1

Wire Resistance Per 1000 Feet

AWG Wire Size	Diam (inches)	Material	Ohms per 1000 ft at 25°C
20	0.032	Copper	10.35
22	0.025	Copper	16.46
24	0.020	Copper	26.17
26	0.016	Copper	41.62
28	0.013	Copper	66.17
30	0.010	Copper	105.2
20	0.032	Nickel	52.78
22	0.025	Nickel	83.95
24	0.020	Nickel	133.47
26	0.016	Nickel	212.26
28	0.013	Nickel	337.47
30	0.010	Nickel	536.52

size nickel wire. We can wind it over a form of some sort. Then we have a unit of a much more convenient size. This is precisely how **wire-wound resistors** are constructed. See Figure 6-2. There is a problem with this type of construction. At radio frequencies, the wire-wound construction causes the resistor to act as an inductor. It does not behave as a pure resistance. For RF circuits, we must have resistors that are noninductive.

There is another way to form a resistor. This is to connect leads to both ends of a block or cylinder of material that has a high resistance. Figure 6-3 shows a resistor made in this manner. It uses a mixture of carbon and clay as the resistive element. This type of resistor is a **carbon-composition resistor**. The proportions of carbon and clay determine the value of the resistor.

An advantage of carbon-composition resistors is the wide range of available values. Other advantages are low inductance and capacitance plus good surge-handling capability. The ability to withstand small power overloads without being completely

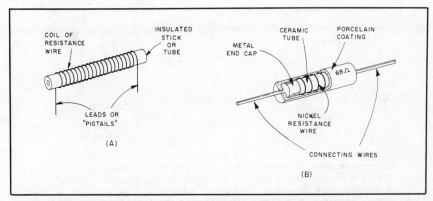

Figure 6-2—A simple wire-wound resistor is shown at A. At B is a commercially produced wire-wound resistor. These resistors are produced by winding resistive wire around a nonconductive form. Most wire-wound resistors are then covered with a protective coating.

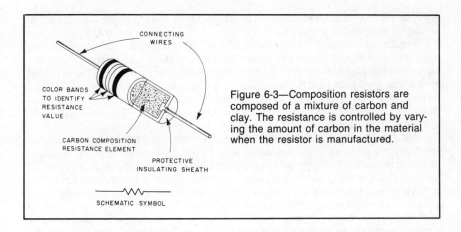

Figure 6-3—Composition resistors are composed of a mixture of carbon and clay. The resistance is controlled by varying the amount of carbon in the material when the resistor is manufactured.

destroyed is another advantage. The main disadvantage is that the resistance of the composition resistor will vary widely. Variations are caused by operating temperature changes and resistor aging.

Another type of carbon resistor is the **carbon-film resistor**. These are manufactured by using high temperatures to break down certain gaseous hydrocarbons. The resulting carbon is then deposited in a thin layer or film on a round ceramic form. The resistor is sealed with a plastic or other insulating material. The thickness of the deposited film provides a means to control the final resistance.

The major advantages of carbon-film resistors are low cost and improved stability with age and temperature changes. These resistors cannot withstand electrical overloads or surges. They can be used as fuses for some applications.

The **metal-film resistor** has replaced the carbon-composition type in many low-power applications today. Metal-film resistors are formed by depositing a thin layer of resistive alloy on a cylindrical ceramic form. Nichrome (an alloy made of nickel and chromium) or other materials are used. See Figure 6-4. This film is then trimmed away in a spiral fashion to form the resistance path. The trimming can be

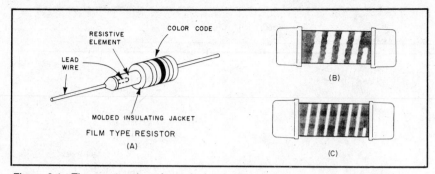

RESISTIVE ELEMENT

COLOR CODE

LEAD WIRE

MOLDED INSULATING JACKET

FILM TYPE RESISTOR

(A)

(B)

(C)

Figure 6-4—The construction of a typical metal-film resistor is shown at A. Film resistors are formed by depositing a thin film of material on a ceramic form. Material is then trimmed off in a spiral to produce the specified resistance for that particular resistor. The excess material may either be trimmed with a mechanical lathe or laser. A lathe produces a rather rough spiral, like that shown in part B. The laser produces a finer cut, shown at C.

done on a mechanical lathe or by using a laser. The resistor is then covered with an insulating material to protect it.

Because of the spiral shape of the resistance path, metal-film resistors will exhibit some inductance. The effects of this inductance increase at higher frequencies. The higher the resistance of the unit, the greater the inductance will be. This is caused by the increased path length. Metal-film resistors provide much better temperature stability than other types. You can easily recognize the "dog bone" shape of a metal-film resistor. They are thinner in the middle than at the ends. This is because of the metal end caps used to connect the leads to the resistive element.

Most resistors have standard fixed values. They are called **fixed resistors**. Variable resistors are sometimes called **potentiometers**. They can be used to adjust the voltage, or potential, in a circuit. Potentiometers are also used a great deal in electronics. The construction of a wire-wound variable resistor and the schematic symbol for all variable resistors is shown in Figure 6-5. Variable resistors are also made with a ring of carbon compound in place of the wire windings. There, a connection is made to each end of the resistance ring. A third contact is attached to a movable arm, or wiper. The wiper can be moved across the ring. As the wiper moves from one end

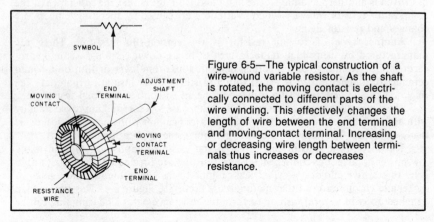

SYMBOL

MOVING CONTACT

END TERMINAL

ADJUSTMENT SHAFT

MOVING CONTACT TERMINAL

END TERMINAL

RESISTANCE WIRE

Figure 6-5—The typical construction of a wire-wound variable resistor. As the shaft is rotated, the moving contact is electrically connected to different parts of the wire winding. This effectively changes the length of wire between the end terminal and moving-contact terminal. Increasing or decreasing wire length between terminals thus increases or decreases resistance.

of the ring to the other, the resistance varies from minimum to maximum.

[Now turn to Chapter 10 and study questions 3AF-1-1.1, 3AF-1-2.1, 3AF-1-2.2, 3AF-1-5.1 and 3AF-1-5.2. Review as needed.]

Color Codes

Standard fixed resistors are usually found in values ranging from 2.7 Ω to 22 MΩ (22 megohms, or 22,000,000 ohms). Resistance tolerances on these standard values can be ±20%, ±10%, ±5%, and ±1%. A tolerance of 10% on a 200-ohm resistor means that the actual resistance of a particular unit may be anywhere from 180 Ω to 220 Ω. (Ten percent of 200 is 20, so the resistance can be 200 + 20 or 200 − 20 ohms.)

It is not practical to print the resistance values on the side of a small resistor. A **color code** shows the value of the resistor. As shown in Figure 6-6, the color of the first three bands shows the resistor value. The color of the fourth band indicates the resistor tolerance.

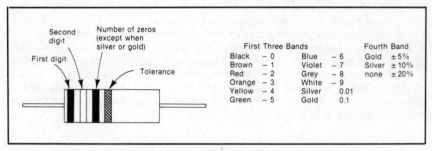

Figure 6-6—Small resistors are labeled with a color code to show their value. For example, proceeding from left to right, a resistor with color bands of yellow, violet, brown, gold is a 470-Ω resistor with a 5% tolerance.

Power Ratings

Resistors are also rated according to the amount of power they can safely handle. Power ratings for wire-wound resistors typically start at 1 watt and range to 10 W or larger. Resistors with higher power ratings are also physically large. Greater surface area is required to dissipate the increased heat generated by large currents. Carbon-composition and film resistors are low-power components. You will find these resistors in 1/8-watt, 1/4-W, 1/2-W, 1-W and 2-W power ratings. A resistor that is being pushed to the limit of its power rating will feel very hot to the touch. Sometimes a resistor gets very hot during normal operation. You should probably replace it with a unit having a higher power-dissipation rating.

[Turn to Chapter 10 and study questions 3AF-1-3.1 through 3AF-1-3.4, 3AF-1-4.1 and 3AF-1-4.2. Review this section as needed.]

INDUCTORS

In Chapter 5 we learned that an inductor stores energy in a magnetic field. When a voltage is first applied to an inductor, with no current flowing through the circuit, the inductor will oppose the current. Remember from Chapter 5 that inductors oppose any change in current. A voltage that opposes the applied voltage is induced in the coil, and this voltage tries to prevent a current. Gradually the current will build up.

Only the resistance in the circuit will limit the current value. The resistance in the wire of the coil will be very small. In the process of getting this current to flow, the coil stores energy. The energy is in the form of a magnetic field around the wire. As the field strength is increasing, the voltage that opposes the current flow is formed. When you shut off the applied current, the magnetic field collapses. The collapsing field returns the stored energy to the circuit. The energy will have the form of a momentary current flowing in the same direction as the original current. This current is produced by the EMF generated by the collapsing magnetic field. Again, this voltage opposes any change in the current already flowing in the coil. The amount of opposition to changes in current is called **reactance**. The reactance of the inductor, the amount of energy stored in the magnetic field and the back EMF induced in the coil all depend on the amount of inductance.

The amount of inductance that a coil exhibits, represented by the symbol L, depends on four things:

    1) the type of material used in the **core**, and its size and location in the coil
    2) the number of turns used to wind the coil
    3) the length of the coil and
    4) the diameter of the coil

Changing any of these factors changes the inductance.

If we add an iron or ferrite core to the coil, the inductance increases. The core of an inductor is the central part of a coil. We can also use brass as a core material in radio-frequency variable inductors. It has the opposite effect of iron or ferrite: The inductance decreases as the brass core enters the coil. A movable core can be placed in a coil, as shown in Figure 6-7. The amount of inductance can be varied as the core moves in and out of the coil. The common schematic symbols for various inductors are shown in Figure 6-8.

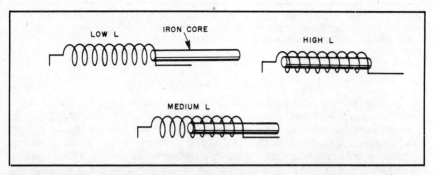

Figure 6-7—The inductance of a coil increases as an iron core is inserted. Many practical inductors of this type use screw threads to allow moving the core in precise amounts when making small changes in inductance.

Ferrite comes from the Latin word for iron. So what is the difference between an iron and a ferrite core? Well, iron cores can be made from iron sheet metal. The layers of material that make up a power transformer core are made that way. Cores can also be made from powdered iron mixed with a bonding material. This holds the powdered iron together so it can be molded into the desired shape. Manufacturers mix other metal alloys with the iron "ferrous" material when they make some cores. We call these ferrite cores. They can select the proper alloy to provide the desired characteristics, producing cores designed to operate best over a specific frequency

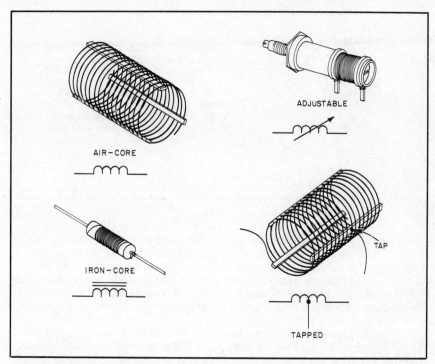

Figure 6-8—Various types of coils, and their schematic symbols. The adjustable inductor uses a movable powdered-iron core or slug. They are sometimes called slug-tuned or permeability-tuned inductors.

range. We use iron cores most often at low frequencies. The audio-frequency range and ac-power circuits are good applications for these cores. Ferrite cores are manufactured in a wide range of compositions. This optimizes their operation for specific radio-frequency ranges.

It takes some amount of energy to magnetize the core material for any inductor. This energy represents a loss in the coil. Winding the coil on an iron material increases the inductance of the coil. The energy needed to magnetize the core also increases. Some coils are wound on thin plastic forms, or made self supporting. Then the only material inside the coil is air. These are air-wound coils. They are among the lowest loss types of inductors. Air-wound coils are normally used in high-power RF circuits where energy loss must be kept to a minimum.

Toroid Cores

A **toroidal inductor** is made by winding a coil of wire on a doughnut-shaped core called a toroid. Toroidal core materials are usually powdered-iron or ferrite compounds. The toroidal inductor is highly efficient. This is because there is no break in the circular core. All the magnetic lines of force remain inside the core. This means there is very little **mutual coupling** between two toroids mounted close to each other in a circuit. See Figure 6-9. Mutual coupling can be a problem, especially in an RF circuit. There, signals from one stage could be coupled into another stage without following the proper signal path. Using toroid cores can often help eliminate the mutual coupling that would occur between other types of coils.

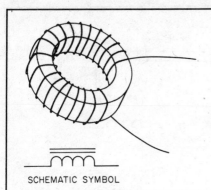

Figure 6-9—A toroidal coil. This type of coil has self-shielding properties. Some inductors in some circuits will interact with the magnetic fields of other inductors. To prevent this a metal shield is used to enclose them. Toroidal inductors confine magnetic fields so well that shields are usually not necessary.

SCHEMATIC SYMBOL

In theory, a toroidal inductor requires no shield to prevent its magnetic field from spreading out and interfering with outside circuits. This self-shielding property also helps keep outside forces from interfering with the toroidal coil. Toroids can be mounted so close to each other that they can almost touch, but because of the way they are made, there will be almost no inductive coupling between them. Toroidal coils wound with only a small amount of wire on ferrite or iron cores can have a very high inductance value.

[Turn to Chapter 10 and study exam questions 3AF-2-1.1 through 3AF-2-1.4, 3AF-2-2.1, 3AF-2-2.2, 3AF-2-3.1 through 3AF-2-3.4 and 3AF-2-4.1 through 3AF-2-4.3. Review any material you have difficulty with before going on.]

CAPACITORS

In Chapter 5 we learned that the basic property of a **capacitor** is the ability to store an electric charge. In an uncharged capacitor, the potential difference between the two plates is zero. As we charge the capacitor, this potential difference increases until it reaches the full applied voltage. The difference in potential creates an **electric field** between the two plates. See Figure 6-10. A field is an invisible force of nature. We put energy into the capacitor by charging it. Until we discharge it, or the charge leaks away somehow, the energy is stored in the electric field. When the field is not moving, we sometimes call it an electrostatic field.

When we apply a dc voltage to a capacitor, current will flow in the circuit until the capacitor is fully charged to the applied voltage. After the capacitor has charged to the full applied voltage, no more current will flow in the circuit. A capacitor opposes changes in voltage. For this reason, capacitors can be used to block dc in a circuit. Because ac voltages are constantly reversing polarity, we can block dc with a capacitor while permitting ac to pass.

Factors That Determine Capacitance

Remember from Chapter 5 that the value of a capacitor is determined by four factors: the area of one side of one plate surface, the distance between the plates, the number of plates and the type of insulating material used between the plates. As the surface area increases, the capacitance increases. More plates also increase the capacitance. The dielectric is the insulating material between the capacitor plates. For a given plate area, as the spacing between the plates decreases, the capacitance increases.

For a given value of plate spacing and area, the type of **dielectric** material used

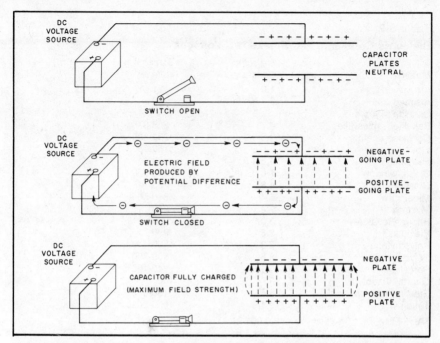

Figure 6-10—Capacitors store energy in an electric field. The final amount of energy stored in a fully-charged capacitor depends on the capacitor value and the applied voltage. If this circuit is used to fully charge the capacitor from a 6-volt battery, the amount of energy stored in the capacitor could be increased by using a 12-volt battery. Another way to increase the amount of stored energy is to use a larger value capacitor.

in the capacitor determines the capacitance. The dielectric is the insulating material between the capacitor plates. A capacitor using air as the dielectric is used for comparison purposes. Other dielectric materials are related to the standard air-dielectric capacitor by the **dielectric constant** of the material used. See Table 6-2. From the chart we can see that if polystyrene is used as the dielectric in a capacitor, it will have 2.6 times the capacitance of an air-dielectric capacitor with the same plate spacing and area.

Voltage Ratings

We cannot increase the voltage applied to a capacitor indefinitely. Eventually we will reach a voltage at which the dielectric material will break down. The breakdown of an insulating material means that the applied voltage is so great that the material will conduct. In a capacitor, this means a spark will jump from one plate to the other. The capacitor will probably be ruined. Different dielectric materials have different **breakdown voltages**, as shown in Table 6-2. The voltage rating of a capacitor is very important. Failure of a capacitor from too much voltage may also damage other components in the circuit. For this reason, capacitors are usually labeled with both their capacitance in microfarads or picofarads and their voltage rating. The voltage rating is specified in working-volts dc (WVDC). Sometimes the tolerance and temperature coefficient are also printed on them.

[Now turn to Chapter 10 and study exam questions 3AF-3-1.1, 3AF-3-1.2, 3AF-3-2.1, 3AF-3-2.2 and 3AF-3-3.1 through 3AF-3-3.3. Review any material you have difficulty with]

Table 6-2

Dielectric Constants and Breakdown Voltages

Material	Dielectric Constant*	Breakdown Voltage**
Air	1.0	21
Alsimag 196	5.7	240
Bakelite™4.4-5.4	300	
Bakelite™, mica filled	4.7	325-375
Cellulose acetate	3.3-3.9	250-600
Fiber	5-7.5	150-180
Formica™	4.6-4.9	450
Glass, window	7.6-8	200-250
Glass, Pyrex™	4.8	335
Mica, ruby	5.4	3800-5600
Mycalex	7.4	250
Paper, Royalgrey	3.0	200
Plexiglas®	2.8	990
Polyethylene	2.3	1200
Polystyrene	2.6	500-700
Porcelain	5.1-5.9	40-100
Quartz, fused	3.8	1000
Steatite, low loss	5.8	150-315
Teflon®	2.1	1000-2000

*At 1 MHz   **In volts per mil (0.001 inch)

Practical Capacitors

Practical capacitors are described by the material used for their dielectric. Mica, ceramic, plastic-film, polystyrene, paper and electrolytic capacitors are in common use today. They each have properties that make them more or less suitable for a particular application.

Mica Capacitors

Mica capacitors consist of many strips of metal foil separated by thin strips of mica. See Figure 6-11. Alternate plates are connected and each set of plates is connected to an electrode. The entire unit is then encased in plastic or ceramic insulating material.

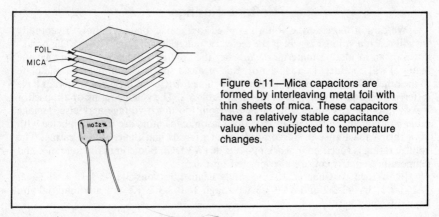

FOIL
MICA

110±2%
EM

Figure 6-11—Mica capacitors are formed by interleaving metal foil with thin sheets of mica. These capacitors have a relatively stable capacitance value when subjected to temperature changes.

An alternative to this form of construction is the "silvered-mica" capacitor. In the silvered-mica type, a thin layer of silver is deposited directly onto one side of the mica. The plates are stacked so that alternate layers of mica are separated by layers of silver.

From Table 6-2 we can see that mica has a very high voltage breakdown rating. For this reason, mica capacitors are frequently used in transmitters and high-power amplifiers. In these applications the ability to withstand high voltages is important. Mica capacitors also have good temperature stability—their capacitance does not change greatly as the temperature changes. Typical capacitance values for mica capacitors range from 1 picofarad to 0.1 microfarad, and voltage ratings as high as 35,000 are possible. Figure 6-12 shows the schematic symbol for all fixed capacitors. Electrolytic capacitors use the same symbol with an added plus mark (+) at the straight side of the symbol.

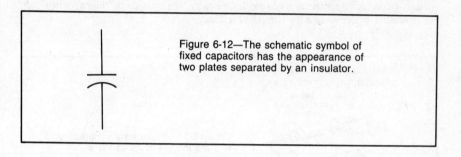

Figure 6-12—The schematic symbol of fixed capacitors has the appearance of two plates separated by an insulator.

Ceramic Capacitors

Ceramic capacitors are constructed by depositing a thin metal film on each side of a ceramic disc, as shown in Figure 6-13. Wire leads are then attached to the metal films. Then the entire unit is covered with a protective plastic or ceramic coating. Ceramic capacitors are inexpensive and easy to construct, and they are in wide use today.

Ordinary ceramic capacitors cannot be used where temperature stability is important—their capacitance will change with a change in temperature. Special ceramic

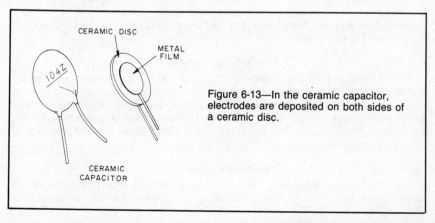

Figure 6-13—In the ceramic capacitor, electrodes are deposited on both sides of a ceramic disc.

capacitors (called NPØ for negative-positive zero) are used for these applications. The capacitance of an NPØ unit will remain substantially the same over a wide temperature range. The range of capacitance values available with ceramic capacitors is typically 1 picofarad to 0.1 microfarad, with working voltages up to 1000.

Ceramic capacitors are often connected across the transformer primary or secondary winding in a power supply. These capacitors, called suppressor capacitors, suppress transient voltage spikes, preventing them from getting through the power supply.

Paper Capacitors

In its simplest form, the **paper capacitor** consists of a layer of paper between two layers of metal foil. The foil and paper are rolled up, as shown in Figure 6-14. Wires are connected to the foil layers, and the capacitor is encased in plastic or dipped in wax to protect it. One end of the capacitor sometimes has a band around it. This does not mean that the capacitor is polarized. It means only that the outside layer of foil is connected to the marked lead. This way, the outside of the capacitor can be grounded where necessary for RF shielding.

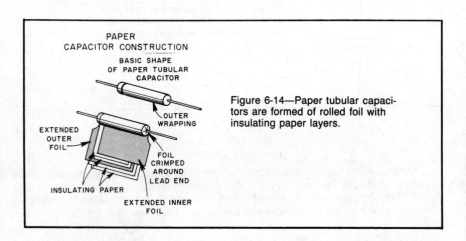

Figure 6-14—Paper tubular capacitors are formed of rolled foil with insulating paper layers.

Paper capacitors can be obtained in capacitance values from about 500 picofarads to around 50 microfarads. Their voltage ratings go up to about 600 WVDC. Paper capacitors are generally inexpensive, but their large size for a given value makes them impractical for some uses.

Plastic-Film Capacitors

Construction techniques similar to those used for paper capacitors are used for **plastic-film capacitors**. Thin sheets of Mylar™ or polystyrene are sandwiched between sheets of metal foil, and rolled into a cylinder. The plastic material gives capacitors with a high voltage rating in a physically small package. Plastic-film capacitors also have good temperature stability. Typical values range from 5 picofarads to 0.47 microfarad.

Electrolytic Capacitors

In **electrolytic capacitors**, the dielectric is formed after the capacitor is manufactured. The construction of aluminum-electrolytic capacitors is very similar to that of paper capacitors. Two sheets of aluminum foil separated by paper are soaked in a chemical solution. Then they are rolled up and placed in a protective casing. After assembly, a voltage is applied to the capacitor. This causes a thin layer of aluminum oxide to form on the surface of the positive plate next to the chemical. The aluminum oxide acts as the dielectric, and the foil plates act as the electrodes.

The layer of oxide dielectric is extremely thin. Electrolytic capacitors can be made with a very high capacitance value in a small package. Electrolytic capacitors are polarized—dc voltages must be connected to the positive and negative capacitor terminals with the correct polarity. The positive and negative electrodes in an electrolytic capacitor are clearly marked. Connecting an electrolytic capacitor incorrectly causes gas to form inside the capacitor and the capacitor may actually explode. This can be very dangerous. At the very least the capacitor will be destroyed by connecting it incorrectly.

Another type of electrolytic capacitor is gaining popularity. Tantalum capacitors have several advantages over aluminum-electrolytic capacitors. Tantalum capacitors can be made even smaller that aluminum-electrolytic capacitors for a given capacitance value. They are manufactured in several forms, including small, water-droplet-shaped solid-electrolyte capacitors. These are formed on a small tantalum pellet that serves as the anode, or positive capacitor plate. An oxide layer on the outside of the tantalum pellet serves as the dielectric. A layer of manganese dioxide is the solid electrolyte. Layers of carbon and silver form the cathode, or negative capacitor plate. The entire unit is dipped in epoxy to form a protective coating on the capacitor. Their characteristic shape explains why these tantalum capacitors are often called "tear drop" capacitors.

Electrolytic capacitors are available in voltage ratings of greater than 400 V, and capacitance values from 1 microfarad to 100,000 microfarads (0.1 farad). Electrolytic capacitors with high capacitance values and/or high voltage ratings are physically very large. Electrolytic capacitors are used in power-supply filters. Large values of capacitance are necessary in these filters to provide good smoothing of the pulsating dc from the rectifier.

Variable Capacitors

In some circuits it is necessary or desirable to vary the capacitance at some point. An example is in the tuning circuit of a receiver or transmitter VFO. We could use a rotary switch to select one of several different fixed-value capacitors. It is much more convenient to use a **variable capacitor**, however. A basic air-dielectric variable capacitor is shown in Figure 6-15. This figure also shows the schematic symbol for a variable capacitor. One set of plates (called the **stator**) is fixed. The other set of plates (called the **rotor**) can be rotated. The rotation controls the amount of plate area shared by the two sets of plates. When the capacitor is fully meshed, all the rotor plates are down in between the stator plates. In this position, the capacitor will have its largest value of capacitance. When the rotor is unmeshed, the value of capacitance is at a minimum. The capacitance can be varied smoothly between the maximum and minimum values.

Another type of variable capacitor is the compression variable. In this type of variable capacitor, the spacing between the two plates is varied. When the spacing is at its minimum, the capacitance is at a maximum value. When the spacing is adjusted to its maximum, the value of capacitance is at a minimum. This type of variable capacitor is usually used as a trimmer capacitor. A trimmer capacitor is a control used to peak or fine tune a part of a circuit. It is usually left alone once adjusted to the correct value.

Variable capacitors are subject to the same rules as other capacitors. If a variable capacitor is to be used with a high voltage on its plates, a large plate spacing or a dielectric with a high breakdown voltage will be required. Air-variable capacitors used

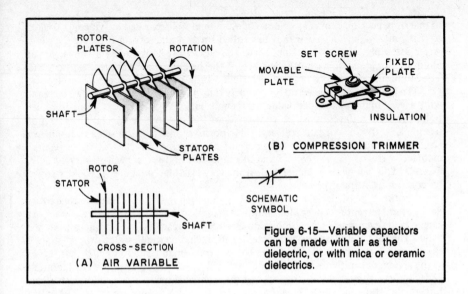

Figure 6-15—Variable capacitors can be made with air as the dielectric, or with mica or ceramic dielectrics.

in transmitters and high-power amplifiers have their plates spaced much farther apart than those used in receivers.

[This completes your study of Chapter 6. Now turn to Chapter 10 and study questions 3AF-3-1.3, 3AF-3-1.4, 3AF-3-2.3, 3AF-3-2.4, 3AF-3-4.1 and 3AF-3-4.2. Review any material you have difficulty with before going on.]

Key Words

Amplifier—A device usually employing electron tubes or transistors to increase the voltage, current, or power of a signal. The amplifying device may use a small signal to control voltage and/or current from an external supply. A larger replica of the small input signal appears at the device output.

Attenuate—To reduce in amplitude.

Band-pass filter—A circuit that allows signals to go through it only if they are within a certain range of frequencies. It attenuates signals above and below this range.

Beat-frequency oscillator (BFO)—An oscillator that provides a signal to the product detector. In the product detector, the BFO signal and the IF signal are mixed to produce an audio signal.

Cutoff frequency—In a high-pass, low-pass, or band pass filter, the cutoff frequency is the frequency at which the filter output is reduced to ½ of the power available at the filter input.

Detector—The stage in a receiver in which the modulation (voice or other information) is recovered from the RF signal.

Direct-conversion receiver—A receiver that converts an RF signal directly to an audio signal with one mixing stage.

Filter—A circuit that will allow some signals to pass through it but will greatly reduce the strength of others.

Frequency modulation—The process of varying the frequency of an RF carrier in response to the instantaneous changes in the modulating signal.

High-pass filter—A filter that allows signals above the cutoff frequency to pass through. It attenuates signals below the cutoff frequency.

Intermediate frequency (IF)—The output frequency of a mixing stage in a super-heterodyne receiver. The subsequent stages in the receiver are tuned for maximum efficiency at the IF.

Low-pass filter—A filter that allows signals below the cutoff frequency to pass through and attenuates signals above the cutoff frequency.

Mixer—A circuit used to combine two or more audio- or radio-frequency signals to produce a different output frequency.

Modulate—To vary the amplitude, frequency, or phase of a radio-frequency signal.

Oscillator—A circuit built by adding positive feedback to an amplifier. It produces an alternating current signal with no input except the dc operating voltages.

Phase modulation—Varying the phase of an RF carrier in response to the instantaneous changes in the modulating signal.

Reactance modulator—A device capable of modulating an ac signal by varying the reactance of a circuit in response to the modulating signal. (The modulating signal may be voice, data, video, or some other kind depending on what type of information is being transmitted.) The circuit capacitance or inductance changes in response to an audio input signal.

Selectivity—A measure of how well a receiver can separate a desired signal from other signals on a nearby frequency.

Sensitivity—The ability of a receiver to detect weak signals.

Stability—A measure of how well a receiver or transmitter will remain on frequency without drifting.

Superheterodyne receiver—A receiver that converts RF signals to an intermediate frequency before detection.

Varactor diode—A component whose capacitance varies as the reverse-bias voltage changes.

Variable-frequency oscillator (VFO)—An oscillator used in receivers and transmitters. The frequency is set by a tuned circuit using capacitors and inductors. The frequency can be changed by adjusting the components in the tuned circuit.

Chapter 7

Practical Circuits

I n this chapter we will discuss low-pass, high-pass and band-pass filter circuits. We'll show you block diagrams of complete transmitters and receivers, and investigate how the stages connect to make them work. You will be directed to turn to Chapter 10 at appropriate points. There you will use the FCC questions as a study aid to review your understanding of the material.

Keep in mind that there have been entire books written on each topic covered in this chapter. You may not understand some of the circuits from our brief discussion. It would be a good idea to consult some other reference books. *The ARRL Handbook for the Radio Amateur* is a good starting point. Even that won't tell you everything about a topic. The discussion in this chapter should help you understand the circuits well enough to pass your Technician class license exam, however.

FILTERS

A problem for hams is harmonic interference to entertainment equipment. Harmonics are multiples of a given frequency. Your transmitter radiates undesired harmonics along with your signal. When transmitting in the high-frequency amateur bands (3.5 to 29.7 MHz), the frequency you're transmitting on is much lower than the TV or FM channels. Some of your harmonics may fall within the home entertainment bands. The entertainment receiver cannot distinguish between the TV or FM signals that are supposed to be there and your harmonics, which are not. If your harmonics are strong enough, they can interfere with the broadcast signal.

Harmonic interference must be cured at your transmitter. It is your responsibility as a licensed amateur. You must see that harmonics from your transmitter are not strong enough to interfere with other services. As mentioned before, all harmonics generated by your transmitter must be attenuated well below the strength of the fundamental frequency. If harmonics from your transmitting equipment exceed these limits, you are at fault.

You can usually tell harmonic interference when you see it on a TV set. This type of interference shows up as crosshatch or a herringbone pattern on the TV screen. See Figure 7-1. Unlike overload, interference from radiated harmonics seldom affects all channels. Rather, it may bother the one channel that is frequency-related to the band you're on. Generally, the low TV channels (channels 2-6) are most affected by harmonics from amateur transmitters operating below 30 MHz. Channels 2 and 6 are especially affected by 10-meter transmitters, and channels 3 and 6 experience trouble from 15-meter transmitters.

There are several possible cures for harmonic interference. We will discuss a few of them here. Try each step in the order that we introduce them. Chances are good that your problem will be quickly solved. You should be familiar with three **filters**:

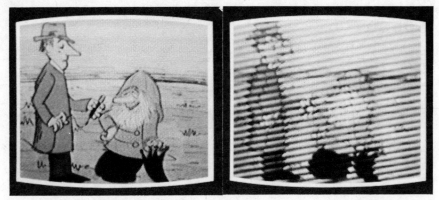

Figure 7-1—On the left is a normal TV picture. Harmonic interference may be visible as crosshatching of the TV screen, as shown on the right. More severe interference can completely destroy the picture. The width of the crosshatch lines will vary, depending on the transmitter frequency.

the low-pass filter, the high-pass filter and the band-pass filter. Filters pass certain frequencies and block others. All modern radio communication devices use filter circuits. These filter circuits allow various kinds of equipment to operate on different frequencies without interfering with each other. Under certain conditions, some equipment may need a little extra "help." A transmitter may need extra filtering of the output signal to reduce interference to television receivers. In other cases, the transmitter may already be well filtered (clean), but a TV receiver may need extra filtering of its input signals. A basic understanding of filters and their applications will often allow you to solve interference problems.

Low-Pass Filters

The first step you should take is installation of a **low-pass filter** in the transmission line between your transmitter and antenna. A low-pass filter is one that passes all frequencies below a certain frequency, called the **cutoff frequency**. We measure the filter cutoff frequency by putting a variable frequency signal into the filter. The input signal power is kept constant while the frequency is increased. At the same time we measure the filter output power. At some frequency the output will begin to decrease. When the output power has decreased to ½ the input power, we have found the cutoff frequency. Frequencies above the cutoff frequency are **attenuated**, or significantly reduced in amplitude. See Figure 7-2.

The cutoff frequency depends on the design of the low-pass filter. The capacitors are chosen to provide a path to ground for the desired high frequencies. The inductor is a value which will pass the low frequencies you want, but will block the higher frequencies. One other important trait that must be considered in the design of a low-pass filter is the characteristic impedance. Most external low-pass filters intended for amateur use will have a characteristic impedance of 50 ohms.

The low-pass filter is always connected between the transmitter and the antenna, as close to the transmitter as possible. The filter should have the same impedance as the feed line connecting the transmitter to the antenna. Filters should be used only in a feed line with a low standing-wave ratio. The cutoff frequency has to be higher than the highest frequency used for transmitting. A filter with a 45-MHz cutoff frequency would be fine for the high-frequency bands. The filter would significantly attenuate 6-meter (50 MHz) signals. Most modern transmitters have a low pass filter built into their output circuitry to prevent excess harmonic radiation.

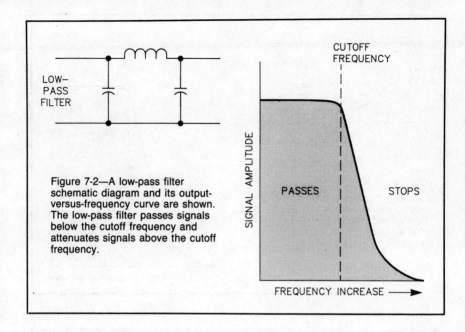

Figure 7-2—A low-pass filter schematic diagram and its output-versus-frequency curve are shown. The low-pass filter passes signals below the cutoff frequency and attenuates signals above the cutoff frequency.

[Now turn to Chapter 10 and study exam questions 3AG-1-1.1, 3AG-1-1.2, 3AG-1-2.1 and 3AG-1-2.2. Review this section if you have trouble with any of these questions.]

High-Pass Filters

Sometimes entertainment devices experience interference from amateur transmissions even though the transmitting device is operating properly. The harmonics can be well below levels necessary to prevent interference. This happens because the design of the entertainment device is inadequate when operating in the presence of strong signals. In this kind of situation, it may be necessary to reduce the level of the amateur signal reaching the entertainment device (TV or FM stereo receiver). The desired higher frequency signals must be allowed to pass unaffected. **High-pass filters** can do this.

A high-pass filter passes all frequencies above the cutoff frequency, and attenuates those below it. See Figure 7-3. A high-pass filter should be connected to a television set, stereo receiver or other home-entertainment device that is being interfered with. It will attenuate the signal from an amateur station. The inductors have a value that allows them to conduct the lower frequencies to ground. The capacitor tends to block these low-frequency signals. At the same time it allows the higher-frequency television or FM broadcast signals to pass through to the receiver. This is useful in reducing the amount of lower-frequency signal (at the ham's operating frequency). The lower frequency signal might overload a television set, causing disruption of reception. For best effect, the filter should be connected as close to the television-set tuner as possible. Put it inside the TV if it is your own set and you don't mind opening it up. You can also attach it directly to the antenna terminals on the back of the television. If the set belongs to your neighbors, have them contact a qualified service technician to install the filter. That way you are not held responsible if something goes wrong with the set later on.

[Turn to Chapter 10 and study questions 3AG-2-1.1, 3AG-2-2.1 and 3AG-2-2.2. Review this section as needed.]

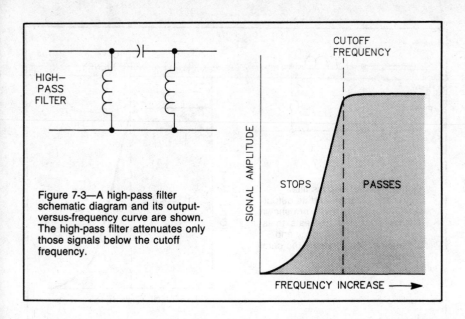

HIGH-PASS FILTER

Figure 7-3—A high-pass filter schematic diagram and its output-versus-frequency curve are shown. The high-pass filter attenuates only those signals below the cutoff frequency.

CUTOFF FREQUENCY

SIGNAL AMPLITUDE

STOPS

PASSES

FREQUENCY INCREASE ⟶

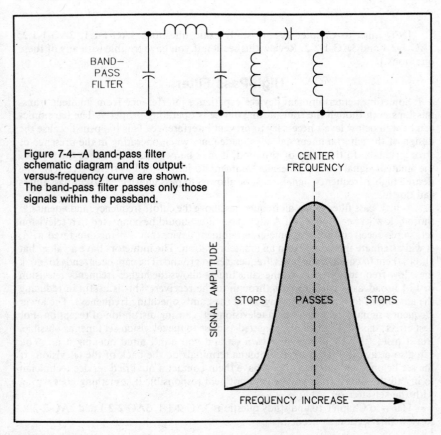

BAND-PASS FILTER

Figure 7-4—A band-pass filter schematic diagram and its output-versus-frequency curve are shown. The band-pass filter passes only those signals within the passband.

CENTER FREQUENCY

SIGNAL AMPLITUDE

STOPS   PASSES   STOPS

FREQUENCY INCREASE ⟶

Band-pass Filters

A **band-pass filter** is a combination of a high-pass and low-pass filter. It passes a desired range of frequencies while rejecting signals above and below the pass band. This is shown in Figure 7-4. Band-pass filters are commonly used in receivers to provide different degrees of rejection. Very narrow filters are used for CW reception. Wider filters are switched in for SSB and AM double-sideband reception. (SSB and AM are explained later.)

[Now turn to Chapter 10 and study questions 3AG-3-1.1, 3AG-3-1.2 and 3AG-3-2.1. Review this section as needed.]

TRANSMITTERS

Amateur transmitters range from the simple to the elaborate. They generally have two basic stages, an **oscillator** and a power **amplifier** (PA). This chapter will introduce the CW transmitter and the FM transmitter.

Amateur receivers can also be simple or complex. You can build a simple solid state receiver that will work surprisingly well. There are plans in *The ARRL Handbook for the Radio Amateur*, including sources for parts and circuit boards. Or, you can buy a sophisticated modern receiver.

Separate receivers and transmitters have been replaced by transceivers in most amateur stations. A transceiver combines circuits necessary for receiving and transmitting in one package. Some circuit sections may perform both transmitting and receiving functions. Other sections are dedicated to transmit-only or receive-only functions. In order to simplify some of the concepts in the following information, we will speak of transmitter or receiver circuits. The principles of operation are the same in transceivers. The transmitter topics apply to transceivers in the transmit mode. Receiver topics apply to transceivers in the receive mode.

Remember that an oscillator produces an ac waveform with no input except the

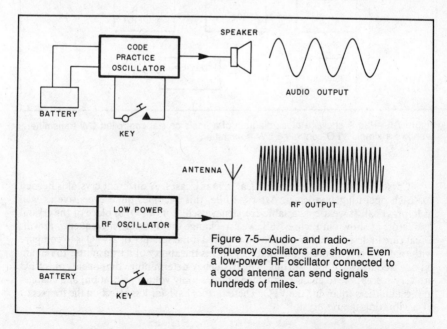

Figure 7-5—Audio- and radio-frequency oscillators are shown. Even a low-power RF oscillator connected to a good antenna can send signals hundreds of miles.

dc operating voltages. You may be familiar with audio oscillators, like a code-practice oscillator. An RF oscillator can be used by itself as a simple low-power transmitter, but the power output is very low. See Figure 7-5. In a practical transmitter the signal from the oscillator is usually fed to one or more amplifier stages.

CW Transmitters

A block diagram of a simple amateur transmitter is shown in Figure 7-6A. It produces CW (continuous-wave unmodulated) radio signals when the key is closed. The simple transmitter consists of a crystal oscillator followed by a driver stage and a power amplifier. A crystal oscillator uses a quartz crystal to keep the frequency of the radio signal constant.

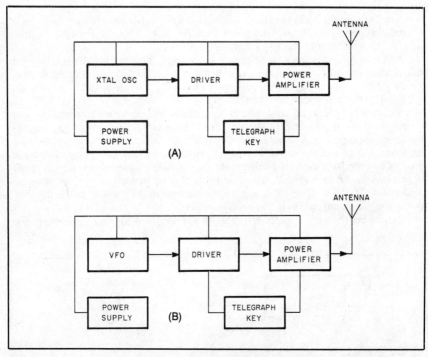

Figure 7-6—Part A shows a block diagram of a basic crystal controlled CW transmitter. B shows a simple VFO controlled CW transmitter.

Crystal oscillators are not practical in many cases. A different crystal is needed for each operating frequency. After a while, this becomes quite expensive as well as impractical. If we use a variable-frequency oscillator (VFO) in place of the crystal oscillator, as shown in Figure 7-6B, we can change the transmitter frequency at will. Great care in design and construction is required if the stability of a VFO is to compare with that of a crystal oscillator. Stability means the ability of a transmitter to remain on one frequency without drifting. The frequency-determining components of a VFO are very susceptible to changes in temperature, supply voltage, vibration, and changes in the amplifiers following the VFO. These factors have far less effect on the frequency of a vibrating quartz crystal.

FM Transmitters

When a radio signal or carrier is modulated, some characteristic of the radio signal is changed in order to convey information. We can transmit information by modulating any property of a carrier. We can **modulate** the frequency or phase of a carrier. **Frequency modulation** and **phase modulation** are closely related. The phase of a signal cannot be varied without also varying the frequency, and vice versa. Phase modulation and frequency modulation are especially suited for channelized local UHF and VHF communication. They feature good audio fidelity and high signal-to-noise ratio.

The simplest type of FM transmitter we could design is shown in Figure 7-7. Remember that a circuit containing capacitance and inductance will be resonant at some frequency. A resonant circuit in the feedback path of an oscillator can control the oscillator frequency. By changing the resonant frequency of the tuned circuit, we can change the oscillator frequency. A capacitor microphone is nothing more than a capacitor with one movable plate (the diaphragm). When you speak into the microphone, the diaphragm vibrates and the spacing between the two plates of the capacitor changes. When the spacing changes, the capacitance value changes. We can connect the microphone to the resonant circuit in our oscillator. Speaking into the microphone will vary the oscillator frequency. We can add frequency multipliers to bring the oscillator frequency up to our operating frequency. Amplifiers to bring up the power will provide a simple FM transmitter.

In practical FM systems, the carrier frequency is varied or modulated by changes in voltage that represent information to be transmitted. This information may originate from a microphone, a computer modem or even a video camera. As the modulating voltage increases and decreases, the carrier frequency will change in proportion (or inverse proportion) to the rise and fall of the modulating voltage. In other words, a modulating voltage that is becoming more positive will increase the carrier frequency. A negative going voltage will decrease the carrier frequency. For some purposes,

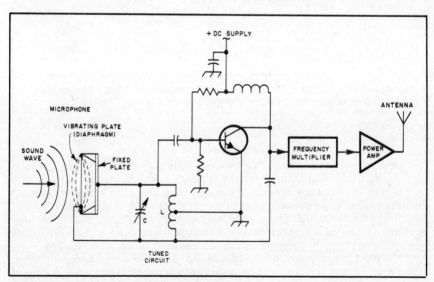

Figure 7-7—This diagram shows a simple FM transmitter. The frequency of the oscillator is changed by changing the capacitance in the resonant circuit. The vibrating plate in the microphone is part of the total capacitance in the resonant circuit.

reversing this relationship will work just as well (FM voice transmitters for instance). The main point is that the carrier frequency is increased and decreased by the modulating signal. The amount of increase or decrease depends on how much the voltage of the modulating signal changes. Speaking loudly into the microphone causes larger variations in carrier frequency (and sometimes distorted audio at the receiving end) than speaking in a normal voice.

One way to shift the frequency of the oscillator would be to use a circuit called a **reactance modulator**. A reactance modulator uses a vacuum tube or transistor. It is wired in a circuit so that it changes either the capacitance or inductance of the oscillator resonant circuit. The changes occur in response to an input signal.

Modern FM transmitters may use a **varactor diode** to modulate the oscillator. A varactor diode is a special diode that changes capacitance when its bias voltage changes. You can connect a varactor diode to a crystal oscillator as shown in Figure 7-8. The varactor diode will "pull" the oscillator frequency slightly when the audio input changes.

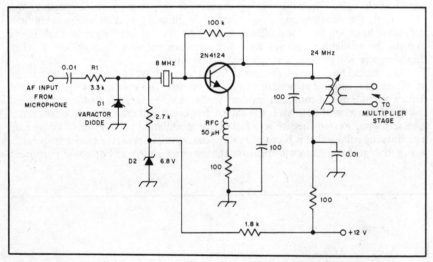

Figure 7-8—A varactor diode (D1) can be used to frequency-modulate an oscillator. The capacitance of the varactor diode changes when the bias voltage varies.

Phase Modulation

One problem with direct-modulated FM transmitters is frequency stability. Designs that allow the oscillator frequency to be easily "moved around" or modulated may have an unfortunate side effect. It becomes more difficult to minimize unwanted frequency shifts arising from changes in temperature, supply voltage, vibration and so on. The frequency multiplying stages also multiply any drift or other instability problems in the oscillator. With phase modulation, the modulation takes place after the oscillator stage. Phase modulation produces what is called indirect FM.

The most common method of generating a phase-modulated telephony signal is to use a reactance modulator. This can be a vacuum tube or a transistor. It is wired in a circuit so that it changes either the capacitance or inductance of a resonant circuit in response to an input signal. The RF carrier is passed through this resonant circuit. Changes in the resonant circuit caused by the reactance modulator effect phase shifts

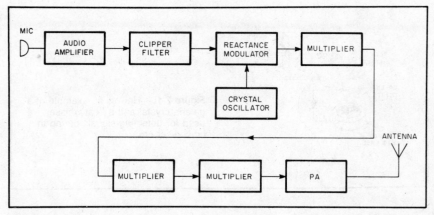

Figure 7-9—This is a block diagram of a phase modulated transmitter, showing the reactance modulator.

in the RF carrier. Figure 7-9 shows the connection of a reactance modulator.

[Now turn to Chapter 10 and study questions 3AG-4-1.2, 3AG-4-1.3, 3AG-4-1.5 and 3AG-4-2.2. Review this section as needed.]

RECEIVERS

The fundamental function of any radio receiver is to change radio-frequency signals (which we can't hear or see) to signals that can be heard (or seen). A good receiver has the ability to find weak radio signals. It will separate them from other signals and interference. Also, it will stay tuned to one frequency without drifting.

The ability of a receiver to detect or discover the presence of weak signals is called **sensitivity**. **Selectivity** is the ability to separate (select) a desired signal from undesired signals. **Stability** is a measure of the ability of a receiver to continue receiving one particular frequency. In general, then, a good receiver is very sensitive, selective and stable.

Detection

The basic duty of a receiver is to extract audio or other information from a radio signal. To detect means to discover. The **detector** stage in a receiver is where the discovery (or recovery) of the information takes place.

The detector is really the heart of a receiver. Everything else is sort of "extra equipment." In the early days of radio, crystal receivers using galena crystals and "cat's whiskers" were in common use. See Figure 7-10. The "cat's whisker" is really just a thin, stiff piece of wire. With the galena crystal it forms a point contact diode. Crystal detectors were an early form of AM (amplitude modulation) detector. AM is generated by varying the amplitude of an RF signal in response to a microphone or other signal source. On today's crowded frequencies, however, more sensitive and selective receivers are required. In addition, crystal receivers cannot be used to receive single sideband (SSB) and CW signals properly. SSB is a special form of AM. Modulation of an RF signal generates new frequencies called sidebands. In SSB transmissions, the original RF signal and one of two side bands are eliminated.

The crystal receiver serves to show how simple a receiver can be. It is an excellent

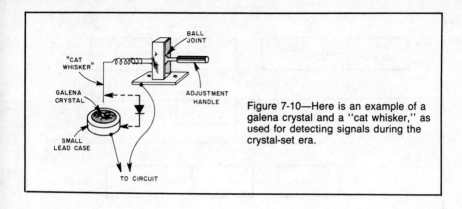

Figure 7-10—Here is an example of a galena crystal and a "cat whisker," as used for detecting signals during the crystal-set era.

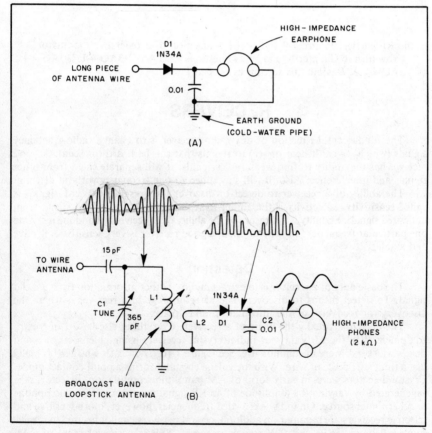

Figure 7-11—Illustration A shows a simple crystal AM receiver. It consists only of a wire antenna, detector diode, capacitor, earphone and earth ground. The circuit at B has a tuned circuit that helps separate the broadcast-band signals, but otherwise operates in the same manner as the circuit at A.

example of how detection is accomplished. Every receiver will have some type of detector stage.

One simple receiver is shown in Figure 7-11A. D1 serves the same function as the early "cat-whisker" detector. Some of these early receivers had circuits similar to Figure 7-11B. This is a tuned-radio-frequency (TRF) receiver. An incoming AM radio signal causes a current to flow from the antenna, through the resonant tuned circuit, to ground. The current induces a voltage with the same waveform in L2. The L1-C1 circuit is resonant at the frequency of the incoming radio signal and tends to reject signals at other frequencies. The diode rectifies the RF signal, allowing only half the waveform to pass through. Capacitor C2 fills in and smoothes out the gaps between the RF cycles of the signal. This leaves only the audio signal to pass on to the headphones. Amplification can be used to improve the sensitivity of this kind of receiver, but other receiver types have much better selectivity.

Direct Conversion Receivers

The next step up in receiver complexity is shown in Figure 7-12. This receiver is called a **direct-conversion receiver**. The incoming signal is combined with a signal from a **variable-frequency oscillator** (VFO) in the **mixer** stage. You should remember how mixing works from your Novice exam. We mix an incoming signal at 7040 kHz with a signal from the VFO at 7041 kHz. The output of the mixer will contain signals at 7040 kHz, 7041 kHz, 14,081 kHz and 1 kHz (the original signals and their sum and difference frequencies). One of these signals (1 kHz) is within the range of human hearing. We use an audio amplifier and hear it in headphones.

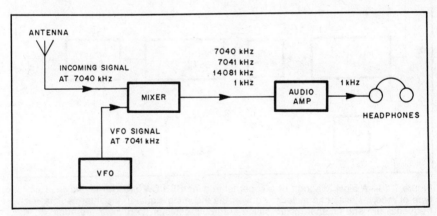

Figure 7-12—A block diagram of a direct-conversion receiver is shown. This type of receiver converts RF signals directly to audio, using only one mixer.

This receiver is called a direct-conversion receiver because the RF signal is converted directly to audio, in one step. By changing the VFO frequency, other signals in the range of the receiver can be converted to audio signals. Direct-conversion receivers are capable of providing good usable reception with relatively simple inexpensive circuits. The major disadvantages of this kind of receiver include problems known as microphonics and ac hum.

Microphonics are the result of using large amounts of audio amplification to build up the mixer (detector) output signal. The audio amplifier in direct-conversion receivers can drive the headphones or speaker by using input signals of a microvolt or less. The problem is, vibrations of circuit components and other parts of the receiver can also generate small voltages. This can produce annoying sounds when the receiver is moved or when controls are adjusted.

The hum problem is related to the 60-Hz ac power distribution system used in buildings and houses. The 60-Hz ac can sometimes modulate the VFO signal. You will then hear a hum or buzz in the headphones.

Superheterodyne Receivers

A block diagram of a simple **superheterodyne receiver** for CW and SSB is shown in Figure 7-13. The first mixer is used to produce a signal at the **intermediate frequency** or IF, typically 9 MHz in a modern receiver. All the amplifiers after this first mixer stage are designed for peak efficiency at the IF. The superhet receiver solves the selectivity/bandwidth problem by converting all signals to the same IF before filtering and amplification. To receive SSB and CW signals, a second mixer, called a product detector, is used. The product detector mixes the IF signal with a signal from the **beat-frequency oscillator (BFO)**. The output from the product detector contains audio components that can be amplified and sent to a speaker or headphones.

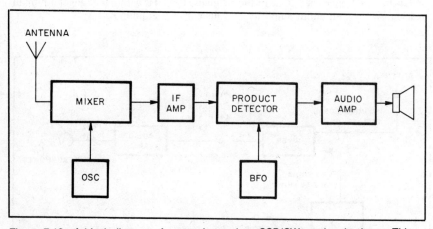

Figure 7-13—A block diagram of a superheterodyne SSB/CW receiver is shown. This type of receiver uses one mixer to convert the incoming signal to the intermediate frequency (IF) and another mixer, called a product detector, to recover the audio or Morse code.

FM Receivers

An FM receiver is organized along the same lines, with a few different stages. It has a wider bandwidth filter and a different type of detector. One type of FM detector is called a frequency discriminator. Most FM receivers have a limiter stage added between the IF amplifier and the detector. FM receivers are designed to be insensitive to amplitude variations caused by impulse-type noise. The limiter chops off noise and amplitude variations from an incoming signal. The frequency dis-

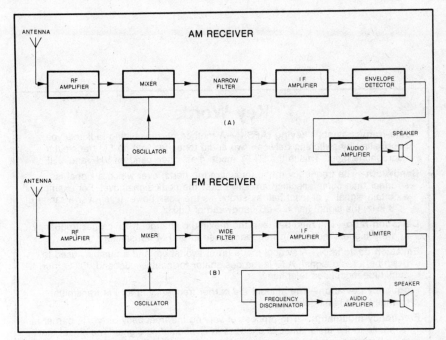

Figure 7-14—AM and FM receivers use similar stages, with a few variations.

criminator produces a waveform varying in amplitude as the frequency of the incoming signal changes. These features make FM popular for mobile communications. A comparison between an AM receiver and an FM receiver is shown in Figure 7-14.

[This completes your study of this chapter. Now turn to Chapter 10 and study questions 3AG-4-1.1, 3AG-4-1.4 and 3AG-4-2.1. Review any material that caused you difficulty before you go on.]

Key Words

Audio-frequency shift keying (AFSK)—A method of transmitting radioteletype information by switching between two audio tones fed into an FM transmitter microphone input. This is the RTTY mode most often used on VHF and UHF.

Bandwidth—The frequency range (measured in hertz) over which a signal is stronger than some specified amount below the peak signal level. For example, if a certain signal is at least half as strong as the peak power level over a range of ± 3 kHz, the signal has a 3-dB bandwidth of 6 kHz.

Deviation ratio—The ratio between the maximum change in RF-carrier frequency and the highest modulating frequency used in an FM transmitter.

Emission designator—A symbol made up of two letters and a number, used to describe a radio signal. A3E is the designator for double-sideband, full-carrier, amplitude-modulated telephony.

Frequency deviation—The amount the carrier frequency in an FM transmitter changes as it is modulated.

Frequency modulation—The process of varying the frequency of an RF carrier in response to the instantaneous changes in a modulating signal. The signal that modulates the carrier frequency may be audio, video, digital data or some other kind of information.

Frequency-shift keying (FSK)—A method of transmitting radioteletype information by switching an RF carrier between two separate frequencies. This is the RTTY mode most often used on the HF amateur bands.

Modulation—The process of varying some characteristic (amplitude, frequency or phase) of an RF carrier for the purpose of conveying information.

Modulation index—The ratio between the maximum carrier frequency deviation and the frequency of the modulating signal at a given instant in an FM transmitter.

Phase—If you consider a point on a waveform, phase is the angular difference between that point and any other point on the same, or another, waveform.

Phase modulation—Varying the phase of an RF carrier in response to the instantaneous changes in the modulating signal.

Reactance modulator—A device capable of modulating an ac signal by varying the reactance of a circuit in response to the modulating signal. (The modulating signal may be voice, data, video, or some other kind depending on what type of information is being transmitted.) The circuit capacitance or inductance changes in response to an audio input signal.

Sidebands—The sum or difference frequencies generated when an RF carrier is mixed with an audio signal.

Splatter—The term used to describe a very wide bandwidth signal, usually caused by overmodulation of a sideband transmitter. Splatter causes interference to adjacent signals.

Chapter 8

Signals and Emissions

T he quality of the signal your station produces reflects on you! You, as a radio amateur and control operator, are responsible (legally and otherwise) for all the signals emanating from your station. Your signals are the way amateurs on the other side of the country or the world get to know you.

One of the most exciting aspects of Amateur Radio is that it offers so many different ways to participate. As a Technician you'll be allowed to use several modes or emission types. This chapter will introduce the modes and emission types you will need to understand for your FCC exam.

MODULATION

Modulation is the process of varying some characteristic (amplitude, frequency or phase) of an RF carrier wave. The carrier varies in accordance with the instantaneous variations of some external signal to convey information. Almost every amateur will think of phone transmission when anyone mentions modulation. The subject of modulation, however, is much broader than that. The principles are the same no matter how the transmitter is modulated. The radio transmission may be modulated by speech or audio tones. The on-off keying of a CW transmitter with Morse code is a form of modulation. The alternating audio tones sent by FM transmitters as digital data is also a form of modulation. Any method of putting information on an RF carrier is a form of modulation.

The sound of your voice consists of physical vibrations in the air, called sound waves. The range of audio frequencies generated by a person's voice may be quite large. The normal range of human hearing is from 20 Hz to 20,000 Hz. All the information necessary to make your voice understood, however, is contained in a narrower band of frequencies. In amateur communications, your audio signals are usually filtered so that only frequencies between about 300 and 3000 hertz are transmitted. This filtering is done to reduce the **bandwidth** of the phone signal.

FCC EMISSION DESIGNATORS

The FCC uses a special system to specify the types of signals (emissions) permitted to amateurs and other users of the radio spectrum. Each **emission designator** has three digits. Table 8-1 shows what each character in the designator stands for. The designators begin with a letter that tells what type of modulation is being used. The second character is a number that describes the signal used to modulate the carrier. The third character specifies the type of information being transmitted.

Table 8-1

Partial List of WARC-79 Emission Designators

1) First Symbol—Modulation Type

Unmodulated carrier	N
Double sideband full carrier	A
Double sideband reduced carrier	R
Single sideband suppressed carrier	J
Vestigial sideband	C
Frequency modulation	F
Phase modulation	G
Various forms of pulse modulation	P, K, L, M, Q, V, W, X

2) Second Symbol—Nature of Modulating Signals

No modulating signal	Ø
A single channel containing quantized or digital information without the use of a modulating subcarrier	1
A single channel containing quantized or digital information with the use of a modulating subcarrier	2
A single channel containing analog information	3
Two or more channels containing quantized or digital information	7
Two or more channels containing analog information	8

3) Third Symbol—Type of Transmitted information

No information transmitted	N
Telegraphy—for aural reception	A
Telegraphy—for automatic reception	B
Facsimile	C
Data transmission, telemetry, telecommand	D
Telephony	E
Television	F

Some of the more common combinations are:

NØN—Unmodulated carrier.

A1A—Morse code telegraphy using amplitude modulation.

F1B—Telegraphy using frequency-shift keying without a modulating audio tone (FSK RTTY). F1B is designed for automatic reception.

F2A—Telegraphy produced by the on-off keying of an audio tone fed into an FM transmitter.

F2B—Telegraphy produced by modulating an FM transmitter with audio tones (AFSK RTTY). F2B is also designed for automatic reception.

F2D—Frequency-modulated data transmission using a modulated subcarrier.

F3E—Frequency-modulated telephony.

G3E—Phase-modulated telephony.

NØN is a radio-frequency signal that stays on continuously with no changes in frequency or amplitude. When heard on an SSB/CW receiver, it sounds like a steady tone. As you tune across the NØN signal, the pitch of the audio tone will change. With AM and FM receivers, no tone or other sounds are heard because there is no modulation. AM and FM receivers will simply become quieter with a lessening of background noise. Because there is no modulation of NØN transmissions, no sidebands are formed.

A1A signals are like NØN signals with the transmitter being turned on and off. In other words, A1A signals are amplitude modulated with only two amplitude levels, on and off. On an SSB/CW receiver, intermittent tones are heard, as in Morse code. AM and FM receivers cannot detect these signals properly. Usually all that is heard are thumping sounds with changes in background noise.

If you wanted to transmit Morse code to people using FM receivers, there are ways to do this. One way is to feed a keyed (on off) audio tone into the microphone input of an FM transmitter. Transmissions of this kind are known as F2A emissions.

RADIOTELEPRINTING

Three types of emissions are commonly used by amateurs in transmitting radio telegraphy for automatic reception: A2B, F1B and F2B. Radio teletype equipment operates with two alternating tones instead of one intermittent tone (as in Morse code). The circuits used to generate these tones for transmitting shift from one frequency to the other, as commanded by the information to be transmitted. These two tones represent digital information. One way to produce the radio teletype signal is to feed two audio tones into the microphone input of an FM transmitter. This method generates F2B emissions. F2B is also known as frequency modulation by **audio-frequency shift keying**. It can be used on the amateur bands above 50.1 MHz.

F2D is a form of data transmission that uses audio frequency-shift keying to frequency modulate an RF carrier. Packet radio, using FM transmitters and receivers, involves reception and transmission of F2D signals.

F2B and F2D are not used on the high-frequency (HF) amateur bands (3.5 to 29.7 MHz). The sidebands generated by these modes would take up far too much space on our more crowded HF bands. As a Technician class licensee, you can transmit F1B signals on 10 meters. An F1B signal shifts back and forth between two distinct, slightly separated frequencies. F1B is also known as **frequency-shift keying**. Some amateur equipment will generate the F1B signal directly. With other amateur equipment, you feed two tones into the microphone input of an SSB transmitter. The transmitter converts this audio frequency signal into a radio-frequency signal. Although this method uses audio-frequency signals to generate F1B signals, it is *not* the same as an F2B emission. The resulting F1B signal shifts between two discrete frequencies. F1B signals occupy a lot less space on the bands. F1B signals do not have wide sidebands (compared to F2B).

If AFSK tones are fed into the microphone input of an amplitude-modulated transmitter, A2B signals will be produced. A2B transmissions are allowed in the amateur bands above 50.1 MHz.

[Now turn to Chapter 10 and study exam questions 3AH-1.1, 3AH-2-1.1, 3AH-2-1.2, 3AH-2-2.1, 3AH-2-2.2, 3AH-2-3.1, 3AH-2-3.2, 3AH-2-4.1, 3AH-2-4.2, 3AH-2-5.1, 3AH-2-5.2, 3AH-2-5.3, 3AH-2-6.1 and 3AH-2-6.2. Review the text material as needed.]

FREQUENCY MODULATION (FM)

We can transmit information by modulating the frequency or phase of a carrier. **Frequency modulation** and phase modulation are closely related. Varying the phase of a signal causes the frequency to vary. Phase modulation and frequency modulation feature good audio fidelity and high signal-to-noise ratio. They are especially suited for channelized local UHF and VHF communication. These benefits are provided as long as the signals are above a minimum level. (When signal levels are below this threshold, other modes such as SSB or CW can give better results.)

In FM systems, when a modulating signal is applied, the carrier frequency increases or decreases in proportion (or inverse proportion) to the instantaneous amplitude of the modulating signal. The *change* in the carrier frequency, or **frequency deviation**, is therefore proportional to the instantaneous amplitude of the modulating signal. Figure 8-1 shows a representation of a frequency modulated signal. The deviation in frequency is slight when the amplitude of the modulating signal is small. Deviation is greatest when the modulating signal reaches its peak. The amplitude of the envelope does not change with modulation. FM receivers are also designed to be insensitive to amplitude variations. This means that an FM receiver is not sensitive to impulse-type noise. This feature makes it popular for mobile communications. The emission designator for a frequency modulated telephony signal is F3E.

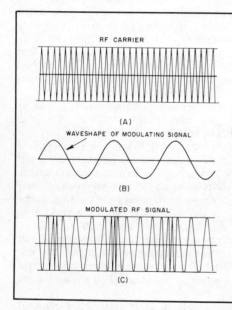

Figure 8-1—Graphical representation of frequency modulation. In the unmodulated carrier at A, each RF cycle takes the same amount of time to complete. When the modulating signal at B is applied, the carrier frequency is increased or decreased according to the amplitude and polarity of the modulating signal.

PHASE MODULATION (PM)

Phase modulation produces what is called indirect FM. As you saw in Chapter 7, the most common method of generating a phase-modulated telephony signal (emission designator G3E) is to use a **reactance modulator**.

The term **phase** essentially means "time," or the time interval between the instant when one thing occurs and the instant when a second related thing takes place. The later event is said to lag the earlier, while the one that occurs first is said to lead. When two waves of the same frequency start their cycles at slightly different times, the time difference (or phase difference) is measured in degrees. (A complete cycle is 360 degrees.) This is shown in Figure 8-2. In phase modulation, we shift the phase of the output wave in response to the audio input signal.

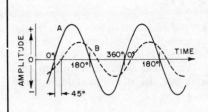

Figure 8-2—When two waves of the same frequency start their cycles at slightly different times, the time difference (or phase difference) is measured in degrees. In this diagram wave B starts 45 degrees (one-eighth cycle) later than wave A, and so lags wave A by 45 degrees. (We can also say that wave A leads wave B by 45 degrees.)

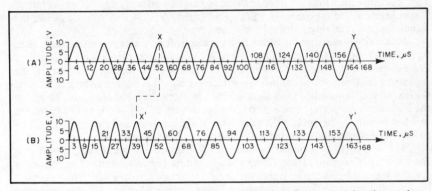

Figure 8-3—Graphical representation of phase modulation. The unmodulated wave is shown at A. After modulation, cycle X′ leads cycle X, and all the cycles to the left of cycle X′ are compressed. To the right, the cycles are spread out.

The output from an RF oscillator will be an unmodulated wave, as shown in Figure 8-3A. The phase-modulated wave is shown in Figure 8-3B. At time 0 and again 168 microseconds later the two waves are in phase. The phase modulation has shifted cycle X′ so that it reaches its peak value in 39 microseconds as compared with 52 microseconds for the unmodulated wave. (Cycle X′ leads cycle X.)

The number of cycles in both waveforms is the same, so the center frequency of the wave has not been changed. All of the cycles to the left of cycle X′ have been compressed, and all the cycles to the right of cycle X′ have been spread out. The frequency of the wave to the left is greater than the center frequency. The frequency to the right is less—the wave has been frequency modulated.

RF CARRIERS

A carrier wave is a constant amplitude unmodulated radio-frequency signal. It has at least one characteristic that can be varied from a known reference by modulation. Even with no modulation, a carrier can provide information about the strength and location of the transmitted signal. Directional antennas and receivers capable of indicating signal strength are used for this purpose.

[Turn to Chapter 10 and study questions 3AH-2-7.1, 3AH-2-7.2, 3AH-2-8.1, 3AH-2-8.2, 3AH-3.1, 3AH-3.2, 3AH-4.1, 3AH-5.1, and 3AH-5.2. Review as needed.]

BANDWIDTH

An important consideration and limitation is the amount of space in the radio-frequency spectrum that a signal occupies. This space is called bandwidth. The bandwidth of a transmission is determined by the information rate. Thus, a pure, continuous, unmodulated carrier (NØN) has a very small bandwidth with no sidebands. A television transmission, which contains a great deal of information, is several megahertz wide.

Receiver Bandwidth

Selectivity is the ability of a receiver to separate two closely spaced signals. Receiver bandwidth determines how well you can receive one signal in the presence of another signal that is very close in frequency. You'll find selectivity to be important. That's because the bands are filled with the signals of a growing number of operators. The enjoyment you'll experience will depend greatly on how well you can isolate the signal you are receiving from all the others nearby.

Selectivity is measured in terms of bandwidth. Bandwidth is nothing more than how wide a range of frequencies is received with the receiver tuned to one frequency. For example, if you can hear signals as much as 3 kHz above and 3 kHz below the frequency to which you are tuned, your receiver must have a bandwidth of at least 6 kHz. If you cannot hear any signals that are more than 200 Hz above or below the frequency to which you are tuned, the bandwidth is only 400 Hz. The narrower this bandwidth is, the greater the selectivity of the receiver, and the easier it will be to copy one signal with another one close by in frequency.

The selectivity of a modern receiver is determined by special filters built into the circuitry. Generally, these filters contain quartz crystals arranged to provide a specific selectivity. Some receivers have several filters built in to enable you to choose between different selectivity bandwidths. This is necessary because different emission types occupy a wider frequency range than others. A 250 Hz bandwidth filter is excellent for separating CW signals on a crowded band, but it's useless for listening to SSB transmissions. A wider filter is needed to allow all the transmitted information to reach the detector.

As a general rule, you should look for a selectivity of 600 Hz or less for CW operation. Many receivers are also designed for single-sideband voice operation, so they come standard with a filter selectivity of around 2.8 kHz. This is usable on CW, but will allow several adjacent CW signals through at the same time. You should try for a receiver with a selectable CW bandwidth of 600 Hz or with provision for adding narrow-bandwidth accessory filters. F1B is a wider emission than A1A (Morse code CW transmissions).

Another selectivity feature available on some receivers is a notch filter. As the name implies, this filter can be used to cut out, or notch, a specific frequency from within the received bandwidth. You will find a notch filter handy when you're trying to receive a signal that is very close in frequency to another signal. By adjusting the notch control, you can effectively eliminate the unwanted signal.

As we have seen in these examples, F1B radio teletype emissions are wider than CW signals and SSB signals are even wider than that. Next, we will consider FM and PM, which can occupy even more bandwidth.

Bandwidth in FM and PM

Frequency deviation is defined as the instantaneous change in frequency for a given signal. We can define the total bandwidth of the signal. The frequency swings just as far in both directions. The total frequency swing is equal to twice the deviation. In addition, there are sidebands that increase the bandwidth still further. A good estimate of the bandwidth is twice the maximum frequency deviation plus the maximum modulating audio frequency:

$$Bw = 2 \times (D + M) \qquad \text{(Equation 8-1)}$$

where

Bw = bandwidth
D = maximum frequency deviation
M = maximum modulating audio frequency

With a transmitter using 5 kHz deviation and a maximum audio frequency of 3 kHz, the total bandwidth is approximately 16 kHz. The actual bandwidth is somewhat greater than this, but it is a good approximation.

Frequency Deviation

With direct FM, frequency deviation is proportional to the amplitude of the modulating signal. With phase modulation, the frequency deviation is proportional to both the amplitude and the frequency of the modulating signal. Frequency deviation is greater for higher audio frequencies. This is actually a benefit, rather than a drawback, of phase modulation. In radio communications applications, speech frequencies between 300 and 3000 Hz need to be reproduced for good intelligibility. In the human voice, however, the natural amplitude of speech sounds between 2000 and 3000 Hz is low. Something must be done to increase the amplitude of these frequencies when a direct FM transmitter is used. A circuit called a preemphasis network amplifies the sounds between 2000 and 3000 Hz. With phase modulation, the preemphasis network is not required, because the deviation already increases with increasing audio input frequency.

Modulation Index

The **sidebands** that occur from FM or PM differ from those resulting from AM. With AM, only a single set of sidebands is produced for each modulating frequency. FM and PM sidebands occur at integral multiples of the modulating frequency on either side of the carrier, as shown in Figure 8-4. Because of these multiple sidebands,

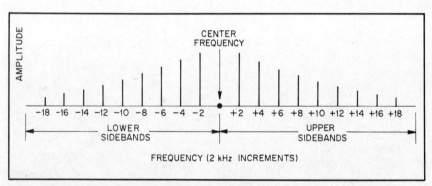

Figure 8-4—When an RF-carrier is frequency-modulated by an audio signal, multiple sidebands are produced. FM sidebands are spaced above and below the center frequency of the carrier. This shows a group of sidebands resulting from the application of a 2-kHz tone to the transmitter modulator. The sidebands farthest from the center frequency are the lowest in amplitude. The sidebands nearest to center frequency can be considered the significant sidebands, their amplitudes being fairly large.

FM or PM signals inherently occupy a greater bandwidth. The additional sidebands depend on the relationship between the modulating frequency and the frequency deviation. The ratio between the peak carrier frequency deviation and the audio modulating frequency is called the **modulation index**.

Given a constant input level to the modulator, in phase modulation, the modulation index is constant regardless of the modulating frequency. For an FM signal, the modulation index varies with the modulating frequency. In an FM system, the ratio of the maximum carrier-frequency deviation to the highest modulating frequency used is called the **deviation ratio**. The modulation index is a variable that depends on a set of operating conditions, but the deviation ratio is a constant. The deviation ratio for narrow-band FM is 5000 Hz (maximum deviation) (divided by 3000 Hz (maximum modulation frequency) or 1.67.

Frequency multiplication offers a means for obtaining any desired amount of frequency deviation, whether or not the modulator is capable of that much deviation. Overdeviation of an FM transmitter causes **splatter**, out-of-channel emissions that can cause interference to adjacent frequencies.

You should understand that a frequency-multiplier stage is an amplifier that produces harmonics of the input signal. By using a filter to select the desired harmonic, the proper output frequency can be obtained. A frequency multiplier is not the same as a mixer stage, which is used to obtain the desired output frequency in an SSB or CW rig. A mixer simply shifts, or translates, an oscillator frequency to some new frequency.

[This completes your study of Chapter 8. Turn to Chapter 10 and study exam questions 3AH-6.1, 3AH-6.2, 3AH-7-1.1, 3AH-7-2.1 and 3AH-7-2.2. Review any material you have difficulty with before going on.]

Key Words

Balanced line—A symmetrical feed line with two conductors having equal but opposite voltages. Neither conductor is at ground potential.

Balun—A transformer used between a BALanced and an UNbalanced system. Used for feeding a balanced antenna with an unbalanced feed line.

Coaxial cable—Feed line with a central conductor surrounded by plastic, foam or gaseous insulation. In turn it is covered by a shielding conductor. The entire cable is usually covered with vinyl insulation.

Cubical quad antenna—An antenna built with its elements in the shape of four-sided loops.

Delta loop antenna—A variation of the cubical quad antenna with triangular elements.

Director—A parasitic element in "front" of the driven element in a multielement antenna.

Driven element—The element connected directly to the feed line in a multielement antenna.

Feed line—The wire or cable used to connect an antenna to the transmitter and receiver.

Front-to-back ratio—The energy radiated from the front of a directional antenna divided by the energy radiated from the back of the antenna.

Gain—An increase in the effective power radiated by an antenna in a certain desired direction. Also, an increase in received signal strength from a certain direction. This is at the expense of power radiated in, or signal strength received from, other directions.

Gamma match—A method of matching coaxial feed line to the driven element of a multielement array.

Half-wavelength dipole antenna—A fundamental antenna one-half wavelength long at the desired operating frequency. It is connected to the feed line at the center. This is a popular amateur antenna.

Horizontally polarized wave—An electromagnetic wave with its electric lines of force parallel to the ground.

Major lobe—The shape or pattern of field strength that points in the direction of maximum radiated power from an antenna.

Parallel-conductor feed line—Feed line constructed of two wires held a constant distance apart. They may be encased in plastic or constructed with insulating spacers placed at intervals along the line.

Parasitic element—Part of a directive antenna that derives energy from mutual coupling with the driven element. Parasitic elements are not connected directly to the feed line.

Polarization—The orientation of the electric lines of force in a radio wave, with respect to the surface of the earth.

Quarter-wavelength vertical antenna—An antenna constructed of a quarter-wavelength-long radiating element placed perpendicular to the earth.

Reflector—A parasitic element placed "behind" the driven element in a directive antenna.

Standing wave ratio—the ratio of maximum voltage to minimum voltage along a feed line. Also the ratio of antenna impedance to feed-line impedance when the antenna is a purely resistive load.

Transmission line—See Feed line.

Unbalanced line—Feed line with one conductor at ground potential, such as coaxial cable.

Vertically polarized wave—A radio wave that has its electric lines of force perpendicular to the surface of the earth.

Yagi antenna—A directive antenna made with a half-wavelength driven element. It has one or more parasitic elements arranged in the same horizontal plane.

Chapter 9

Antennas and Transmission Lines

Y ou need an antenna system to radiate a signal from your transmitter. An antenna system includes a number of items. The antenna itself (what is actually radiating the signal), and the **feed line** are the most important parts. Any coupling devices or matching networks are also included. These devices transfer power from the transmitter to the line and from the line to the antenna. In some simple systems the feed line may be part of the antenna. Coupling devices might not be used there. Technician class license applicants should know the basics of antennas and transmission lines commonly used in Amateur Radio.

THE HALF-WAVELENGTH DIPOLE ANTENNA

A popular amateur antenna is approximately half the wavelength of the transmitted signal. This antenna is the basis of many more complex forms of antennas. This basic antenna is called the **half-wavelength dipole antenna**.

The wavelength of a radio wave is related to the speed of light and the frequency of the wave by the equation:

$$\lambda = \frac{c}{f} \qquad \text{(Equation 9-1)}$$

where
  λ = wavelength in meters
  c = the speed of light in meters per second
  f = the frequency in hertz

The speed of light in a vacuum is 3×10^8 meters per second, and if we use the frequency in megahertz (hertz $\times 10^6$) the equation reduces to:

$$\lambda = \frac{3 \times 10^2}{f(MHz)} = \frac{300}{f(MHz)} \qquad \text{(Equation 9-2)}$$

For example, using Equation 9-2 to find the wavelength of a 7.15-MHz signal:

$$\lambda = \frac{300}{f(MHz)} = \frac{300}{7.15} = 42.0 \text{ meters}$$

A half wavelength in meters is given by the formula:

$$\frac{\lambda}{2} = \frac{150}{f(MHz)} \qquad \text{(Equation 9-3)}$$

The length of a dipole antenna will be somewhat less than this. One reason is because radio waves propagate more slowly along metal than in a vacuum. The capacitance present at the ends of the dipole have an even larger effect than the velocity of propagation. This makes it necessary to shorten the dipole even more. If we convert from metric to US Customary units and allow for shortening effects, we can derive an approximate relation between frequency (f) and length (in feet) for a wire antenna:

$$L \text{ (ft)} = \frac{468}{f(\text{MHz})} \qquad \text{(Equation 9-4)}$$

The length of a dipole for 7.15 MHz is:

$$L \text{ (ft)} = \frac{468}{7.15} = 65.5 \text{ ft}$$

PARASITIC BEAM ANTENNAS

The radiation pattern of the half-wavelength dipole antenna is described as bidirectional. It radiates equally well in two directions. A unidirectional pattern provides maximum radiation in one direction. Unidirectional radiation can be produced by coupling the half-wavelength antenna to additional elements. The additional elements may be made of wire or metal tubing.

Parasitic Excitation

In most multiple-element antennas, the additional elements are not directly connected to the feed line. They receive power by mutual coupling from the **driven element**. The driven element is the element connected to the feed line. The additional elements then reradiate the power in the proper phase relationship. Proper phasing achieves **gain** or directivity over a simple half-wavelength dipole. These elements are called **parasitic elements**.

There are two types of parasitic elements. A **director** is generally shorter than the driven element. A director is located at the "front" of the antenna. A **reflector** is generally longer than the driven element. A reflector is located at the "back" of the antenna. See Figure 9-1. The direction of maximum radiation from a parasitic beam antenna is from the reflector through the driven element to the director. The term **major lobe** refers to the region of maximum radiation from a directional antenna. The major lobe is also sometimes referred to as the main lobe. Communication in

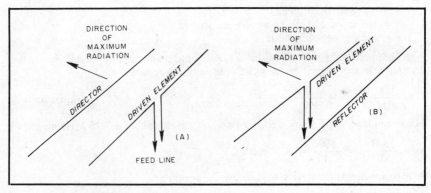

Figure 9-1—In a directional antenna, the reflector element is placed "behind" the driven element. The director goes in "front" of the driven element.

different directions may be achieved by rotating the array in the azimuthal, or horizontal, plane.

Yagi Arrays

The Yagi arrays in Figure 9-2 are examples of antennas that make use of parasitic elements to produce a unidirectional radiation pattern. A Yagi antenna has at least two elements. One element is a driven element. The other elements are directors and/or reflectors. These elements are usually parallel to each other and made of straight metal tubing. Though typical HF antennas of this type have three elements, some may have six or more. Multiband **Yagi antennas** have many elements. Some elements work on some frequencies and others are used for different frequencies. The radiation pattern of the Yagi antenna is shown in Figure 9-3. This pattern indicates that the antenna will reject signals coming from the sides and back. It selects mainly those signals from a desired direction. There are several types of Yagi antennas.

There are many different methods of connecting the feed line to the driven element of a Yagi antenna. The most common feed system, shown in Figure 9-4, is called the gamma match.

The length of the driven element in the most common type of Yagi antenna is approximately an electrical one-half wavelength. This means that Yagi antennas are most often used for the 20-meter band and higher frequencies. At frequencies below 14 MHz, the physical dimensions become very large. Those antennas require special construction techniques, heavy-duty supporting towers and large rotators. Yagis for 40 meters are fairly common. Rotatable Yagis for 80 meters are few and far between. You can overcome the mechanical difficulties of large Yagis. Some amateurs build

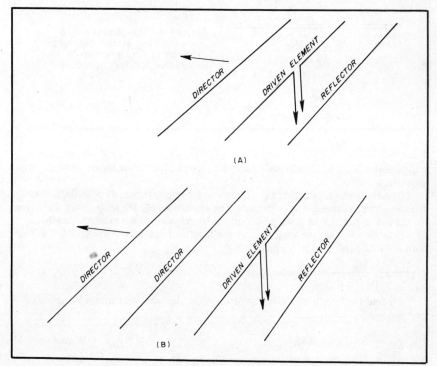

Figure 9-2—At A, a three-element Yagi. At B, a four-element Yagi with two directors.

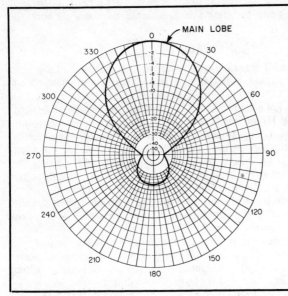

Figure 9-3—The directive pattern for a typical three-element Yagi. Note that this pattern is essentially unidirectional, with most of the radiation in the direction of the major lobe.

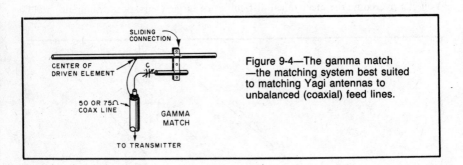

Figure 9-4—The gamma match —the matching system best suited to matching Yagi antennas to unbalanced (coaxial) feed lines.

nonrotatable Yagi antennas for 40 and 80 meters. The elements are made of wire supported on both ends.

In a three-element beam, the director will be approximately 5% (0.05) shorter than the driven element. The reflector will be approximately 5% longer than the driven element. We can derive a set of equations to calculate these element lengths.

Equation 9-4 can be used to find the length of the driven element for a Yagi with a 21.12-MHz center frequency.

$$L \text{ (ft)} = \frac{468}{f\text{(MHz)}} = \frac{468}{21.12} = 22.16 \text{ ft}$$

To find the length of a director or reflector element, we can use Equations 9-5 and 9-6:

$$L_{director} = L_{driven} \times 0.95 \hspace{3cm} \text{(Equation 9-5)}$$
$$L_{director} = 22.16 \text{ ft} \times 0.95$$
$$L_{director} = 21.05 \text{ ft}$$

$$L_{reflector} = L_{driven} \times 1.05 \qquad \text{(Equation 9-6)}$$
$$L_{reflector} = 22.16 \text{ ft} \times 1.05$$
$$L_{reflector} = 23.27 \text{ ft}$$

There can be considerable variation of these lengths, however. The actual lengths depend on the spacing between elements and the diameter of the elements. Whether the elements are made from tapered or cylindrical tubing also makes a difference.

The **polarization** of the signal from a Yagi antenna is determined by antenna placement relative to the surface of the earth. Yagi elements are usually parallel to the surface of the earth, as shown in Figure 9-5A. This way the transmitted wave will be horizontally polarized. If the elements are perpendicular to the surface of the earth, as in Figure 9-5B, the wave will have vertical polarization.

[Now turn to Chapter 10 and study questions 3AI-1-1.1 through 3AI-1-1.7. Review the material in this section as needed.]

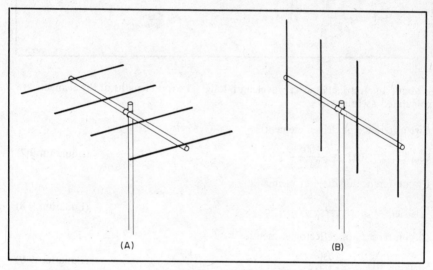

Figure 9-5—The orientation of the elements in a Yagi antenna determines the polarization of the transmitted wave. If the elements are horizontal as shown at A, the wave will have horizontal polarization. Vertical mounting, shown at B, produces a vertically polarized wave.

Cubical Quad Antennas

The **cubical quad antenna** also uses parasitic elements. This antenna is sometimes simply called a quad. The two elements of the quad antenna are usually wire loops. The total length of the wire in the driven element is approximately one electrical wavelength. A typical quad, shown in Figure 9-6, has two elements—a driven element and a reflector. A two-element quad could also use a driven element with a director. You can add more elements, such as a reflector and one or more directors. Additional elements can be added to increase the directivity of the antenna. By installing additional loops of the proper dimensions, we can work different frequency bands with the same antenna. This is commonly done as a modification to a 20-meter quad to provide 15- and 10-meter coverage. The elements of the quad are usually square. Each side

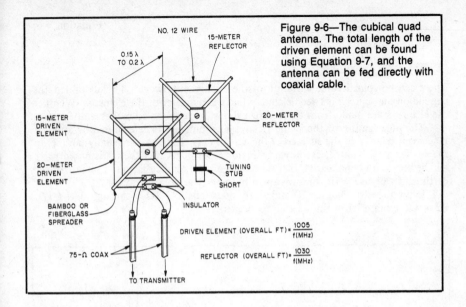

Figure 9-6—The cubical quad antenna. The total length of the driven element can be found using Equation 9-7, and the antenna can be fed directly with coaxial cable.

NO. 12 WIRE
15-METER REFLECTOR
0.15 λ TO 0.2 λ
15-METER DRIVEN ELEMENT
20-METER REFLECTOR
20-METER DRIVEN ELEMENT
TUNING STUB
SHORT
BAMBOO OR FIBERGLASS SPREADER
INSULATOR
DRIVEN ELEMENT (OVERALL FT) = $\frac{1005}{f(MHz)}$
REFLECTOR (OVERALL FT) = $\frac{1030}{f(MHz)}$
75-Ω COAX
TO TRANSMITTER

is about an electrical quarter wavelength long. The total lengths of the elements are calculated as follows:

Circumference of driven element:

$$C_{\text{driven element}}(ft) = \frac{1005}{f(MHz)}$$ (Equation 9-7)

Circumference of director element:

$$C_{\text{director}}(ft) = \frac{975}{f(MHz)}$$ (Equation 9-8)

Circumference of reflector element:

$$C_{\text{reflector}}(ft) = \frac{1030}{f(MHz)}$$ (Equation 9-9)

So for a 28.2-MHz cubical quad, the element circumference would be:

$$C_{\text{driven element}}(ft) = \frac{1005}{28.2} = 35.6 \text{ ft}$$

$$C_{\text{director}}(ft) = \frac{975}{28.2} = 34.6 \text{ ft}$$

$$C_{\text{reflector}}(ft) = \frac{1030}{28.2} = 36.5 \text{ ft}$$

Remember that these equations give the total length of the elements. To find the length of each side of the antenna, we must divide the total length by 4.

The polarization of the signal from a quad antenna can be changed. Polarization is determined by where the feed point is located on the driven element. See Figure 9-7. The feed point can be located in the center of a horizontal side, parallel to the earth's surface. Then the transmitted wave will be a **horizontally polarized wave**. A

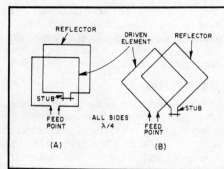

Figure 9-7—The feed point of a quad antenna determines the polarization. Fed in the middle of the bottom side (as shown at A) or at the bottom corner (as shown at B), the antenna produces a horizontally polarized wave. Vertical polarization is produced by feeding the antenna at the side in either case.

vertically polarized wave results when the antenna is fed in the center of a vertical side. We can turn the antenna 45 degrees, so it looks like a diamond. When the antenna is fed at the bottom corner, the transmitted wave will be horizontally polarized. If the antenna is fed at a side corner, the transmitted wave will be vertically polarized.

Polarization is especially important when building antennas for use at VHF and UHF. Propagation at these frequencies is generally line of sight. The polarization of a VHF or UHF signal does not change from transmitting antenna to receiving antenna. Best signal reception will occur when both transmitting and receiving stations use the same polarization. The polarization of an HF signal may change many times as it passes through the ionosphere. Antenna polarization is not as important there.

Delta Loop Antennas

The **delta loop antenna**, shown in Figure 9-8, is a variation on the quad. In a

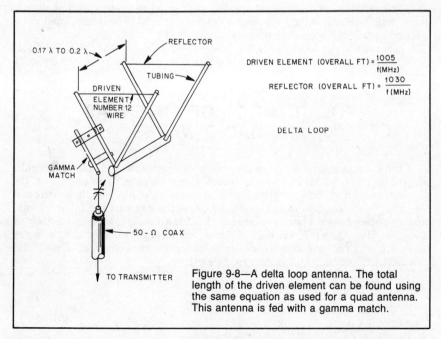

$$\text{DRIVEN ELEMENT (OVERALL FT)} = \frac{1005}{f(\text{MHz})}$$

$$\text{REFLECTOR (OVERALL FT)} = \frac{1030}{f(\text{MHz})}$$

DELTA LOOP

Figure 9-8—A delta loop antenna. The total length of the driven element can be found using the same equation as used for a quad antenna. This antenna is fed with a gamma match.

delta loop, the elements are triangular rather than square. The same equations used for the quad will work to calculate the element lengths for the delta loop. Divide the total length by 3 to find the length of each side of the delta loop antenna. The total length is divided by 3 because of the triangular shape of the elements. The radiation pattern of a typical quad is similar to that of the Yagi shown in Figure 9-3.

Front-To-Back Ratio

You can find the **front-to-back ratio** of an antenna. Divide the power radiated in the direction of maximum radiation by the power radiated in the reverse direction. The direction of maximum radiation is the front of the antenna. Point your antenna directly at a receiving station and transmit. Turn the antenna 180 degrees and transmit again. The difference in the received signal strength is the front-to-back ratio. Front-to-back ratio is usually expressed in decibels.

One way to increase the front-to-back ratio is by adjusting the length of the parasitic elements. Another way is to adjust their position relative to the driven element and each other. The tuning condition that gives maximum attenuation to the rear is considerably more critical than the condition for maximum forward gain.

Gain versus Spacing

When we speak of the **gain** of an antenna, we mean gain referenced to another antenna, used for comparison. One common reference antenna is an ideal half-wavelength dipole in free space. The gain of an antenna with parasitic elements varies with the boom length of the antenna. The spacing between elements, and the length and number of parasitic elements also affect gain.

The maximum front-to-back ratio seldom, if ever, occurs under the same conditions that yield maximum forward gain. It is frequently necessary to sacrifice some forward gain to get the greatest front-to-back ratio, or vice versa. Changing the tuning and spacing of the elements affects the impedance of the driven element and the bandwidth of the antenna. The most effective method to increase the bandwidth of a parasitic beam antenna is to use larger-diameter elements.

Theoretically, an optimized three-element Yagi antenna has a gain of slightly more than 7 dB over a half-wavelength dipole. A typical two-element quad has a gain of about 6.5 dB over a half-wavelength dipole.

[Turn to Chapter 10 and study questions 3AI-1-2.1 through 3AI-1-2.4, 3AI-1-3.1, 3AI-2-1.2, 3AI-2-1.3, 3AI-2-2.4 and 3AI-2-2.5. Review any material you had difficulty with.]

POLARIZATION OF ANTENNAS AND RADIO WAVES

An electromagnetic wave consists of moving electric and magnetic fields. Remember that a field is an invisible force of nature. We can't see radio waves; the best we can do is to show a representation of where the energy is in the electric and magnetic fields. We did this in Chapter 5 to show the magnetic flux around a coil, and the electric field in a capacitor. We can visualize a traveling radio wave as looking something like Figure 9-9. The lines of electric and magnetic force are at right angles to each other. They are also perpendicular to the direction of travel. These fields can have any position with respect to the earth.

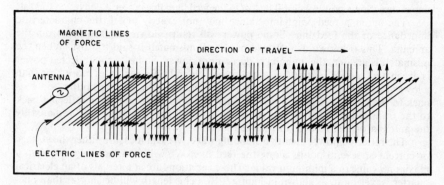

Figure 9-9—A horizontally polarized electromagnetic wave. The electrical lines of force are parallel to the ground. The fields are stronger where the arrows are closer together. The points of maximum field strength correspond to the points of maximum amplitude for the voltage producing the RF signal.

Horizontal Polarization

Polarization is defined by the direction of the electric lines of force in a radio wave. A wave with electric lines of force that are parallel to the surface of the earth is a horizontally polarized wave.

Vertical Polarization

The electric lines of force can also be perpendicular to the earth. This wave is said to be vertically polarized. The polarization of a radio wave as it leaves the antenna is determined by the orientation of the antenna. For example, a half-wavelength dipole parallel to the surface of the earth transmits a wave that is horizontally polarized. A half-wavelength dipole perpendicular to the surface of the earth transmits a vertically polarized wave. An amateur mobile whip antenna is mounted vertically on an automobile. It transmits a wave that is vertically polarized.

A popular vertical antenna is the **quarter-wavelength vertical antenna**. It is often used to obtain low-angle radiation when a beam or dipole cannot be placed far enough above ground. Low-angle radiation refers to signals that travel closer to the horizon, rather than signals that are high above the horizon. Low-angle radiation is usually advantageous when you are attempting to contact distant stations. Vertical antennas of any length radiate vertically polarized waves.

Most man-made noise tends to be vertically polarized. Thus, a horizontally polarized antenna will receive less noise of this type than a vertical antenna will.

[Now turn to Chapter 10 and study exam questions 3AI-2-1.1, 3AI-2-2.1, 3AI-2-2.2 and 3AI-2-2.3. Review as needed.]

STANDING-WAVE RATIO (SWR)

In a perfect antenna system, all the power put into the feed line by the transmitter would be radiated by the antenna. The power traveling from the transmitter to the antenna is called forward power. A practical antenna system will behave most like this ideal when the feed-line impedance is matched to the antenna feed-point impedance. Some power will still be lost in the feed line; this power is converted to heat.

The antenna feed-point impedance may not exactly match the characteristic impedance of the feed line. Some power will return down the feed line from the antenna. This is known as reflected power. A mismatch is said to exist. When the mismatch is greater, the antenna reflects more power. Some of the reflected power is dissipated as heat. A transmission line with high losses generates more heat than a lower-loss line. The reflected power is reflected again by the transmitter and goes back to the antenna. There, some power will be radiated, and some reflected back to the transmitter. This process continues until all the power is either radiated by the antenna or lost as heat in the feed line.

The reflected power creates a standing wave. We can measure either the voltage or current at several points along the feed line. We will find that it varies from a maximum value to a minimum value. These variations are at intervals of an electrical quarter wavelength, as shown in Figure 9-10. (This length will be shorter than one-quarter wavelength in free space. The type of transmission line used affects the distance between points in Figure 9-10.) The standing wave is created when energy reflected from the antenna meets the forward energy from the transmitter. The two waves alternately cancel and reinforce each other.

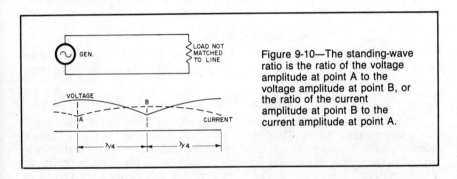

Figure 9-10—The standing-wave ratio is the ratio of the voltage amplitude at point A to the voltage amplitude at point B, or the ratio of the current amplitude at point B to the current amplitude at point A.

The **standing-wave ratio** or SWR is defined as the ratio of the maximum voltage to the minimum voltage in the standing wave:

$$SWR = \frac{E_{max}}{E_{min}} \qquad \text{(Equation 9-10)}$$

An SWR of 1:1 means you have no reflected power. The voltage and current are constant at any point along the feed line. The line is said to be "flat." If the load is completely resistive (no reactance), then the SWR can be calculated. Divide the line characteristic impedance by the load resistance, or vice versa. Use whichever gives a value greater than one:

$$SWR = \frac{Z_0}{R} \text{ or } SWR = \frac{R}{Z_0} \qquad \text{(Equation 9-11)}$$

where
Z_0 = characteristic impedance of the transmission line
R = load resistance (not reactance)

For example, if you feed a 100-ohm antenna with 50-ohm transmission line, the SWR is 100 / 50 or 2:1. Similarly, if the impedance of the antenna is 25 ohms the SWR is 50 / 25 or 2:1.

When a high standing-wave ratio condition exists, losses in the feed line are increased. This is because of the multiple reflections from the antenna and transmitter. Each time the transmitted power has to travel up and down the feed line, a little more energy is lost as heat. This effect is not as great as some people believe, however. Some line losses are less than 2 dB (such as for 100 feet of RG-8 or RG-58 cables up to about 30 MHz). The SWR would have to be greater than 3:1 to add an extra decibel of loss because of the SWR. (A high SWR will cause the output power to drop drastically with many solid-state transceivers. The drop in power is caused by internal circuits that sense the high SWR and automatically reduce output power to protect the transceiver.)

A **transmission line** should be terminated in a resistance equal to its characteristic impedance. Then maximum power is delivered to the antenna, and transmission line losses are minimized. This would be an ideal condition. Such a perfect match is seldom realized in a practical antenna system.

BALANCED AND UNBALANCED CONDITIONS

Feed Lines

Two basic types of feed line are in common use at amateur stations. **Coaxial cable** has a center conductor surrounded by insulation. There is a shield braid around that, and plastic insulation over the whole thing. See Figure 9-11. The second type is **parallel conductor feed line**. It has two conductors spaced some distance apart, with plastic insulation between them. Ladder line or open-wire line is a type of parallel-conductor feed line. It is made from bare wires with plastic spacers every few inches. These spacers maintain the separation between the two conductors. See Figure 9-12. The distance between the conductors determines the characteristic impedance of the feed line. For 300-ohm line the conductors are about ½ inch apart. If the spacing is closer to 1 inch, the characteristic impedance is about 450 ohms. Open wire line is best for operation at a high SWR.

Coaxial cable is an **unbalanced line**. The shield braid is connected to ground, while the center conductor is not. Twin lead is a **balanced line**; neither conductor is connected to ground. Most antennas are inherently balanced devices. For example, there are equal currents and voltages in both halves of a dipole antenna. Balanced antennas should ideally be fed with balanced feed line.

Antennas

A balanced antenna is symmetrical about the feed point. A half-wavelength dipole is an example of a balanced antenna. Neither side of a dipole connects to ground.

An unbalanced antenna is not symmetrical about its feed point. A ¼-wavelength vertical is an example of an unbalanced antenna. One side of a vertical antenna (the radials) often connects to ground.

Baluns

The **balun** is a special kind of transformer. When a balanced antenna is fed with unbalanced line, a device called a balun can be used. The balun (short for BALanced to UNbalanced) is used to transform the unbalanced condition on the feed line to the balanced condition at the antenna. Some baluns can also be used to match impedances between the antenna and transmission line.

When a coaxial cable is used to feed a half-wave dipole, a balun can be used to decouple the transmission line from the antenna. This helps prevent the feed line from radiating a signal, which could distort the antenna radiation pattern.

It is important to note that even though a transmitter may "see" a pure 50-ohm load, and the SWR between the transmitter and matching device is 1:1, the SWR measured in the transmission line between the matching device and the antenna may

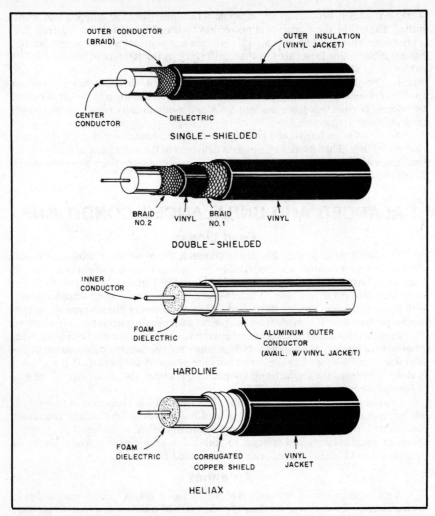

Figure 9-11—Some of the common types of coaxial-cable feed line, showing cable construction.

not be 1:1. Sometimes you want to use SWR as a measure of the impedance match between your antenna and feed line. In that case, the best place to measure SWR is at the antenna feed point. Some SWR meters have a remote sensing unit, so the actual meter can be located some distance away for convenience in reading the meter.

Inserting a matching device between the transmission line and the transmitter does not affect the SWR in the transmission line between the matching device and the antenna; the matching device only provides a 50-ohm resistive load for the transmitter. The SWR meter should be placed between the transmitter and the Transmatch (adjustable matching network) to adjust the Transmatch properly.

[Now turn to Chapter 10 and study questions 3AI-3-3.1 through 3AI-3-3.3, 3AI-4-1.1, 3AI-4-1.2, 3AI-4-2.1, 3AI-4-2.2, 3AI-4-3.1 and 3AI-4-3.2. Review as needed.]

FEED-LINE ATTENUATION

A transmission line (or feed line) is used to transfer power from the transmitter to the antenna. The two basic types of transmission lines generally used, parallel-conductor feed line and coaxial cable, can be constructed in a variety of forms. Both types can be divided into two classes: those in which the majority of the space between the conductors is air, and those in which the conductors are embedded in and separated by a solid plastic or foam insulation (dielectric).

Air-Insulated Feed Lines

Air-insulated transmission lines have the lowest loss per unit length (usually expressed in dB/100 ft). Adding a solid dielectric between the conductors increases the losses in the feed line. The power loss causes heating of the dielectric. As frequency increases, conductor and dielectric losses become greater for coaxial and parallel conductor feed lines.

A typical type of construction used for parallel-conductor or "ladder line" air-insulated transmission lines is shown in Figure 9-12. The two wires are supported a fixed distance apart by means of insulating rods called spacers. Spacers are commonly made from phenolic, polystyrene, isolantite or Lucite™. The spacers generally vary in length from 1 to 6 inches. The shorter lengths are desirable at the higher frequencies so the conductors are held a small fraction of a wavelength apart and radiation from the transmission line will be minimized. Spacers are placed along the line at intervals that are small enough to prevent the two lines from moving appreciably with respect to each other. This type of line is sometimes referred to as "open-wire" feed line. An advantage of this type of transmission line is that it can be operated at a high SWR and still retain its low-loss properties.

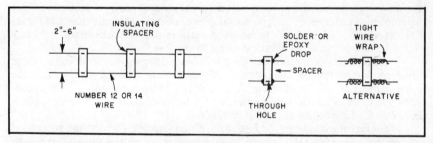

Figure 9-12—"Ladder line" is constructed of two parallel wires separated at intervals by insulators.

The characteristic impedance of an air-insulated parallel-conductor line depends on the diameter of the wires used in the feed line and the spacing between them. The greater the spacing between the conductors, the higher the characteristic impedance of the feed line. The impedance decreases, however, as the size of the conductors increases. The characteristic impedance of a feed line is not affected by the length of the line.

Solid-Dielectric Feed Lines

Transmission lines in which the conductors are separated by a flexible dielectric have several advantages over air-dielectric line: They are less bulky, maintain more uniform spacing between conductors, are generally easier to install and are neater in appearance. Both parallel-conductor and coaxial lines are available with this type of insulation.

One disadvantage of these types of lines is that the power loss per unit length is greater than air-insulated lines because of the dielectric. As the frequency increases, the dielectric losses become greater. The power loss causes heating of the dielectric. Under conditions of high power or high SWR, the dielectric may actually melt and cause short circuits or arcing inside the line.

One common parallel-conductor feed line has a characteristic impedance of 300 ohms. This type of line is used for TV antenna feed line, and is usually called twin lead. Twin lead consists of two number 20 wires that are molded into the edges of a polyethylene ribbon about a half-inch wide. The presence of the solid dielectric lowers the characteristic impedance of the line as compared to the same conductors in air.

The fact that part of the field between the conductors exists outside the solid dielectric leads to operating disadvantages. Dirt or moisture on the surface of the ribbon tends to change the characteristic impedance. Weather effects can be minimized, however, by coating the feed line with silicone grease or car wax. In any case, the changes in the impedance will not be very serious if the line is terminated in its characteristic impedance (Z_0). If there is a considerable standing-wave ratio, however, then small changes in Z_0 may cause wide fluctuations of the input impedance.

Coaxial Cable

Coaxial cables are available in flexible and semiflexible formats, but the fundamental design is the same for all types. Some coaxial cables have stranded-wire center conductors, while others employ a solid copper conductor. Coaxial cables commonly used by radio amateurs have a characteristic impedance of 50 ohms or 75 ohms. The outer conductor or shield may be a single layer of copper braid, a double layer of braid (more effective shielding), or solid copper or aluminum (most effective shielding). An outer insulating jacket (usually vinyl) provides protection from dirt, moisture and chemicals. Typical examples of coaxial cables are shown in Figure 9-11. Exposure of the dielectric to moisture and chemicals will cause it to deteriorate over time and increase the electrical losses in the line.

The characteristic impedance of a coaxial line depends on the diameter of the center conductor, the dielectric constant of the insulation between conductors, the inside diameter of the shield braid and the distance between conductors. The characteristic impedance of coaxial cables increases for larger-diameter shield braids, but it decreases for larger-diameter center conductors.

The larger the diameter, the higher the power capability of the line because of the increased dielectric thickness and conductor size. In general, losses decrease as the cable diameter increases, because there is less power lost in the conductor.

Amateurs commonly use RG-8, RG-58 and RG-174 coaxial cable. RG-8 has the least loss and RG-174 has the highest losses. As the frequency increases, the conductor and dielectric losses become greater, causing more attenuation of the signal in the cable. Although the cost, size and weight are larger for RG-8, it is usually the best choice (of the cables mentioned) for a run of over 150 feet for frequencies up to 54 MHz (the amateur 6-meter band). Open-wire transmission line has the least loss of any feed line type commonly used by amateurs.

Of the common coaxial cables discussed here, RG-8 has the least loss. RG-58 has a bit more loss than RG-8, and RG-174 has the most loss of these cables. RG-174 is normally used for cables that connect sections of a transmitter or receiver, or for short interconnecting cables in a low-power system.

Extra cable length increases attenuation. When using coaxial cable, you should try to use a matched antenna and feed line. You should then be able to change feed line lengths without significantly affecting the antenna system. Then your feed line has to be only long enough to reach your antenna. A low SWR on the line means that the impedance "seen" by the transmitter will be about the same regardless of line length. You can cut off or shorten excess cable length to reduce attenuation of the signal caused by feed-line loss. (This does not apply to multiple antennas in phased arrays or line sections used for impedance-matching purposes.)

Attenuation is not affected by the characteristic impedance of a matched line if the spacing between the conductors in the coaxial cable is a small fraction of a wavelength at the operating frequency.

[Now turn to Chapter 10 and study examination questions 3AI-5-1.1, 3AI-5-1.2, 3AI-5-2.1, 3AI-5-2.2 and 3AI-5-3.1 through 3AI-5-3.3.]

RF SAFETY

Chapter 4 listed some safety precautions to follow when you are working with radio-frequency energy. There are two areas you should be especially concerned with. When your antenna system operates with a high SWR or at high power levels, there can be high RF voltages on the feed line. Antennas can develop high RF voltages even with a low-SWR transmission line. High RF voltages present a burn hazard to anyone who comes in contact with bare conductors of a feed line or antenna. Bare conductors of open-wire feed line should be positioned where people will not come in contact with it when you are transmitting. When possible, insulated antenna feed lines should also be located to prevent accidental contact. This will prevent electrical shocks and RF burns.

Make sure to position your antenna so that people or animals cannot touch it. This will minimize the risk of RF burns as well as exposure to radio-frequency energy. You will usually install your antennas as high as you can get them, anyway. Ground-mounted antennas do require some extra thought. You can surround them with a wooden fence or some other protective device. Position such antennas away from sidewalks and other high traffic areas when possible. This will minimize human exposure to radio-frequency energy from your antenna.

[Now complete your study of this chapter by turning to Chapter 10 and studying examination questions 3AI-5-6.1 and 3AI-6-2.1. Review this section as needed. That completes your study for the Technician class examination. Be sure you have also studied the FCC Rules and Regulations and can answer all of the questions in Subelement 3AA of Chapter 10.]

Chapter 10

Element 3A Question Pool

Don't Start Here!

B efore you read the questions and answers printed in this chapter, be sure to read the appropriate text in the previous chapters. Use these questions as review exercises when suggested in the text. You should not attempt to memorize all 326 questions and answers in the Element 3A question pool. The material presented in this book has been carefully written and scientifically prepared to guide you step by step through the learning process. By understanding the electronics principles and Amateur Radio concepts as they are presented, your insight into our hobby and your appreciation for the privileges granted by an Amateur Radio license will be greatly enhanced.

This chapter contains the complete question pool for the Technician class license written exam (Element 3A). The FCC specifies the minimum number of questions for an Element 3A exam, and also specifies that a certain number of questions from each subelement must appear on the exam. Most VECs give 25-question Element 3A exams. There must be at least five questions from the Rules and Regulations section, subelement 3AA, three questions from the Operating Procedures section, subelement 3AB, and so on. The number of questions to be selected from each section is printed at the beginning of each subelement, and is summarized in Table 10-1.

Table 10-1
Technician Exam Content

Subelement	3AA	3AB	3AC	3AD	3AE	3AF	3AG	3AH	3AI
Number of Questions	5	3	3	4	2	2	1	2	3

The FCC now allows Volunteer-Examiner teams to select the questions that will be used on amateur exams. If your test is coordinated by the ARRL/VEC, however, your test will be prepared by the VEC. The questions with multiple choice answers and distractors printed here were released by the Volunteer Examiner Coordinators' Question Pool Committee for use between November 1, 1989 and October 31, 1992. Most VECs have agreed to use these multiple-choice answers and distractors. If your exam is coordinated by the ARRL/VEC or one of the other VECs using these multiple-choice answers, they will appear on your exam exactly as they are printed here. Some VECs may use the questions printed here with different answers and/or distractors;

check with the VEC coordinating your test session.

We have listed page references along with the answers in the answer key section of this chapter. These page numbers indicate where you will find the text discussion related to each question. If you have problems with a question, refer back to the page listed with the answer. You may have to study beyond the listed page numbers. There are no page references for the questions in the Rules and Regulations subelement (3AA). Reference material for the rules and regulations questions can be found in *The FCC Rule Book*, published by the ARRL.

You should be aware that the Federal Communications Commission reorganized its rules governing Amateur Radio effective September 1, 1989. The new "Part 97" of the FCC Rules is contained in the Eighth (or later) Edition of *The FCC Rule Book,* published by ARRL. This question pool was developed before the new version of the rules was released. These new rules may affect the answers to some questions. VECs and Volunteer Examiners should use some discretion when selecting questions for an exam. The VEC Question Pool Committee is expected to release a supplement to this Element 3A (Technician) Question Pool on or before March 1, 1990. This supplement will include revisions to questions and answers affected by the new rules. The supplement is expected to be put into use on July 1, 1990.

Please remember that if you do not have an Amateur Radio license, you will also have to pass the Novice license exam, Element 2, to earn a Technician class license.

SUBELEMENT 3AA—Commission's Rules (5 Exam Questions)

3AA-1.1　What is the *control point* of an amateur station?
A. The operating position of an Amateur Radio station where the control operator function is performed
B. The operating position of any Amateur Radio station operating as a repeater user station
C. The physical location of any Amateur Radio transmitter, even if it is operated by radio link from some other location
D. The variable frequency oscillator (VFO) of the transmitter

3AA-1.2　What is the term for the operating position of an amateur station where the control operator function is performed?
A. The operating desk
B. The control point
C. The station location
D. The manual control location

3AA-2.1　What are the HF privileges authorized to a Technician control operator?
A. 3700 to 3750 kHz, 7100 to 7150 kHz (7050 to 7075 kHz when terrestrial station location is in Alaska or Hawaii or outside Region 2), 14,100 to 14,150 kHz, 21,100 to 21,150 kHz, and 28,100 to 28,150 kHz only
B. 3700 to 3750 kHz, 7100 to 7150 kHz (7050 to 7075 kHz when terrestrial station location is in Alaska or Hawaii or outside Region 2), 21,100 to 21,200 kHz, and 28,100 to 28,500 kHz only
C. 28,000 to 29,700 kHz only
D. 3700 to 3750 kHz, 7100 to 7150 kHz (7050 to 7075 kHz when terrestrial station location is in Alaska or Hawaii or outside Region 2), and 21,100 to 21,200 kHz only

3AA-2.2　Which operator licenses authorize privileges on 52.525 MHz?
A. Extra, Advanced only
B. Extra, Advanced, General only
C. Extra, Advanced, General, Technician only
D. Extra, Advanced, General, Technician, Novice

3AA-2.3　Which operator licenses authorize privileges on 146.52 MHz?
A. Extra, Advanced, General, Technician, Novice
B. Extra, Advanced, General, Technician only
C. Extra, Advanced, General only
D. Extra, Advanced only

3AA-2.4　Which operator licenses authorize privileges on 223.50 MHz?
A. Extra, Advanced, General, Technician, Novice
B. Extra, Advanced, General, Technician only
C. Extra, Advanced, General only
D. Extra, Advanced only

3AA-2.5　Which operator licenses authorize privileges on 446.0 MHz?
A. Extra, Advanced, General, Technician, Novice
B. Extra, Advanced, General, Technician only
C. Extra, Advanced, General only
D. Extra, Advanced only

3AA-3.1 How often do Amateur Radio operator and station licenses need to be renewed?
A. Every 10 years
B. Every 5 years
C. Every 2 years
D. They are lifetime licenses

3AA-3.2 The FCC currently issues amateur licenses carrying 10-year terms. What is the "grace period" during which the FCC will renew an expired 10-year license?
A. 2 years
B. 5 years
C. 10 years
D. There is no grace period

3AA-3.3 How do you modify an Amateur Radio operator and station license?
A. Properly fill out FCC Form 610 and send it to the FCC in Gettysburg, Pa
B. Properly fill out FCC Form 610 and send it to the nearest FCC field office
C. Write the FCC at their nearest field office
D. There is no need to modify an amateur license between renewals

3AA-4.1 On what frequencies within the 6-meter band may emission F3E be transmitted?
A. 50.0-54.0 MHz only
B. 50.1-54.0 MHz only
C. 51.0-54.0 MHz only
D. 52.0-54.0 MHz only

3AA-4.2 On what frequencies within the 2-meter band may emission F3F be transmitted?
A. 144.1-148.0 MHz only
B. 146.0-148.0 MHz only
C. 144.0-148.0 MHz only
D. 146.0-147.0 MHz only

3AA-4.3 What emission mode may always be used for station identification, regardless of the transmitting frequency?
A. A1A
B. F1B
C. A2B
D. A3E

3AA-5.1 What is the nearest to the band edge the transmitting frequency should be set?
A. 3 kHz for single sideband and 1 kHz for CW
B. 1 kHz for single sideband and 3 kHz for CW
C. 1.5 kHz for single sideband and 0.05 kHz for CW
D. As near as the operator desires, providing that no sideband, harmonic, or spurious emission (in excess of that legally permitted) falls outside the band

3AA-5.2 When selecting the transmitting frequency, what allowance should be made for sideband emissions resulting from keying or modulation?

A. The sidebands must be adjacent to the authorized Amateur Radio frequency band in use

B. The sidebands must be harmonically related frequencies that fall outside of the Amateur Radio frequency band in use

C. The sidebands must be confined within the authorized Amateur Radio frequency band occupied by the carrier

D. The sidebands must fall outside of the Amateur Radio frequency band in use so as to prevent interference to other Amateur Radio stations

3AA-6-1.1 FCC Rules specify the maximum transmitter power that you may use with your Amateur Radio station. At what point in your station is the transmitter power measured?

A. By measuring the final amplifier supply voltage inside the transmitter or amplifier

B. By measuring the final amplifier supply current inside the transmitter or amplifier

C. At the antenna terminals of the transmitter or amplifier

D. On the antenna itself, after the feed line

3AA-6-1.2 What is the term used to define the average power during one radio-frequency cycle at the crest of the modulation envelope?

A. Peak transmitter power

B. Peak output power

C. Average radio-frequency power

D. Peak envelope power

3AA-6-2.1 Notwithstanding the numerical limitations in the FCC Rules, how much transmitting power shall be used by an amateur station?

A. There is no regulation other than the numerical limits

B. The minimum power level required to achieve S9 signal reports

C. The minimum power necessary to carry out the desired communication

D. The maximum power available, as long as it is under the allowable limit

3AA-6-3.1 What is the maximum transmitting power permitted an amateur station on 146.52 MHz?

A. 200 watts PEP output

B. 500 watts ERP

C. 1000 watts dc input

D. 1500 watts PEP output

3AA-6-4.1 What is the maximum transmitting power permitted an amateur station in beacon operation?

A. 10 watts PEP output

B. 100 watts PEP output

C. 500 watts PEP output

D. 1500 watts PEP output

3AA-7-1.1　What is the maximum sending speed permitted for an emission F1B transmission between 28 and 50 MHz?
A. 56 kilobauds
B. 19.6 kilobauds
C. 1200 bauds
D. 300 bauds

3AA-7-1.2　What is the maximum sending speed permitted for an emission F1B transmission between 50 and 220 MHz?
A. 56 kilobauds
B. 19.6 kilobauds
C. 1200 bauds
D. 300 bauds

3AA-7-1.3　What is the maximum sending speed permitted for an emission F1B transmission above 220 MHz?
A. 300 bauds
B. 1200 bauds
C. 19.6 kilobauds
D. 56 kilobauds

3AA-7-2.1　What is the maximum frequency shift permitted for emission F1B when transmitted below 50 MHz?
A. 100 Hz
B. 500 Hz
C. 1000 Hz
D. 5000 Hz

3AA-7-2.2　What is the maximum frequency shift permitted for emission F1B when transmitted above 50 MHz?
A. 100 Hz or the sending speed, in bauds, whichever is greater
B. 500 Hz or the sending speed, in bauds, whichever is greater
C. 1000 Hz or the sending speed, in bauds, whichever is greater
D. 5000 Hz or the sending speed, in bauds, whichever is greater

3AA-7-3.1　What is the maximum bandwidth permitted an amateur station transmission between 50 and 220 MHz using a non-standard digital code?
A. 20 kHz
B. 50 kHz
C. 80 kHz
D. 100 kHz

3AA-7-3.2　What is the maximum bandwidth permitted an amateur station transmission between 220 and 902 MHz using a non-standard digital code?
A. 20 kHz
B. 50 kHz
C. 80 kHz
D. 100 kHz

3AA-7-3.3　What is the maximum bandwidth permitted an amateur station transmission above 902 MHz using a non-standard digital code?
A. 20 kHz
B. 100 kHz
C. 200 kHz, as defined by Section 97.66 (g)
D. Any bandwidth, providing that the emission is in accordance with section 97.63 (b) and 97.73 (c)

3AA-8-1.1 How must a newly upgraded Technician control operator with a *Certificate of Successful Completion of Examination* identify the station while it is transmitting on 146.34 MHz pending receipt of a new operator license?

A. The new Technician may not operate on 146.34 until his or her new license arrives

B. The licensee gives his or her call sign, followed by the word "temporary" and the identifier code shown on the certificate of successful completion

C. No special form of identification is needed

D. The licensee gives his or her call sign and states the location of the VE examination where he or she obtained the certificate of successful completion

3AA-8-2.1 Which language(s) must be used when making the station identification by telephony?

A. The language being used for the contact may be used if it is not English, providing the US has a third-party traffic agreement with that country

B. English must be used for identification

C. Any language may be used, if the country which uses that language is a member of the International Telecommunication Union

D. The language being used for the contact must be used for identification purposes

3AA-8-3.1 What aid does the FCC recommend to assist in station identification when using telephony?

A. A speech compressor

B. Q signals

C. An internationally recognized phonetic alphabet

D. Distinctive phonetics, made up by the operator and easy to remember

3AA-9-1.1 What is the term used to describe a one-way radio communication conducted in order to facilitate measurement of radio equipment characteristics, adjustment of radio equipment or observation of propagation phenomena?

A. Beacon operation

B. Repeater operation

C. Auxiliary operation

D. Radio control operation

3AA-9-2.1 What class of Amateur Radio operator license must you hold to operate a beacon station?

A. Technician, General, Advanced or Amateur Extra class

B. General, Advanced or Amateur Extra class

C. Amateur Extra class only

D. Any license class

3AA-10.1 What is the maximum mean output power an amateur station is permitted in order to operate under the special rules for radio control of remote model craft and vehicles?

A. One watt

B. One milliwatt

C. Two watts

D. Three watts

3AA-10.2 What information must be indicated on the writing affixed to the transmitter in order to operate under the special rules for radio control of remote model craft and vehicles?

- A. Station call sign
- B. Station call sign and operating times
- C. Station call sign and licensee's name and address
- D. Station call sign, class of license, and operating times

3AA-10.3 What are the station identification requirements for an amateur station operated under the special rules for radio control of remote model craft and vehicles?

- A. Once every ten minutes, and at the beginning and end of each transmission
- B. Once every ten minutes
- C. At the beginning and end of each transmission
- D. Station identification is not required

3AA-10.4 Where must the writing indicating the station call sign and the licensee's name and address be affixed in order to operate under the special rules for radio control of remote model craft and vehicles?

- A. It must be in the operator's possession
- B. It must be affixed to the transmitter
- C. It must be affixed to the craft or vehicle
- D. It must be filed with the nearest FCC Field Office

3AA-11-1.1 What is an amateur *emergency communication*?

- A. An Amateur Radio communication directly relating to the immediate safety of life of individuals or the immediate protection of property
- B. A communication with the manufacturer of the amateur's equipment in case of equipment failure
- C. The only type of communication allowed in the Amateur Radio Service
- D. A communication that must be left to the Public Safety Radio Services; for example, police and fire officials

3AA-11-1.2 What is the term for an Amateur Radio communication directly related to the immediate safety of life of an individual?

- A. Immediate safety communication
- B. Emergency communication
- C. Third-party communication
- D. Individual communication

3AA-11-1.3 What is the term for an Amateur Radio communication directly related to the immediate protection of property?

- A. Emergency communication
- B. Immediate communication
- C. Property communication
- D. Priority traffic

3AA-11-2.1 Under what circumstances does the FCC declare that a *general state of communications emergency* exists?

- A. When a declaration of war is received from Congress
- B. When the maximum usable frequency goes above 28 MHz
- C. When communications facilities in Washington, DC, are disrupted
- D. In the event of an emergency disrupting normally available communication facilities in any widespread area(s)

3AA-11-2.2   How does an amateur operator request the FCC to declare that a *general state of communications emergency* exists?
A. Communication with the FCC Engineer-In-Charge of the affected area
B. Communication with the US senator or congressman for the area affected
C. Communication with the local Emergency Coordinator
D. Communication with the Chief of the FCC Private Radio Bureau

3AA-11-2.3   What type of instructions are included in an FCC declaration of a *general state of communications emergency*?
A. Designation of the areas affected and of organizations authorized to use radio communications in the affected area
B. Designation of amateur frequency bands for use only by amateurs participating in emergency communications in the affected area, and complete suspension of Novice operating privileges for the duration of the emergency
C. Designation of the areas affected and specification of the amateur frequency bands or segments of such bands for use only by amateurs participating in emergency communication within or with such affected area(s)
D. Suspension of amateur rules regarding station identification and business communication

3AA-11-2.4   During an FCC-declared *general state of communications emergency*, how must the operation by, and with, amateur stations in the area concerned be conducted?
A. All transmissions within all designated amateur communications bands other than communications relating directly to relief work, emergency service, or the establishment and maintenance of efficient Amateur Radio networks for such communications shall be suspended
B. Operations shall be governed by part 97.93 of the FCC rules pertaining to emergency communications
C. No amateur operation is permitted in the area during the duration of the declared emergency
D. Operation by and with amateur stations in the area concerned shall be conducted in the manner the amateur concerned believes most effective to the speedy resolution of the emergency situation

3AA-12.1   What is meant by the term *broadcasting*?
A. The dissemination of radio communications intended to be received by the public directly or by intermediary relay stations
B. Retransmission by automatic means of programs or signals emanating from any class of station other than amateur
C. The transmission of any one-way radio communication, regardless of purpose or content
D. Any one-way or two-way radio communication involving more than two stations

3AA-12.2   What classes of station may be automatically retransmitted by an amateur station?
A. FCC licensed commercial stations
B. Federally or state-authorized Civil Defense stations
C. Amateur Radio stations
D. National Weather Service bulletin stations

3AA-12.3   Under what circumstances, if any, may a broadcast station retransmit the signals from an amateur station?
A. Under no circumstances
B. When the amateur station is not used for any activity directly related to program production or newsgathering for broadcast purposes
C. If the station rebroadcasting the signal feels that such action would benefit the public
D. When no other forms of communication exist

3AA-12.4   Under what circumstances, if any, may an amateur station retransmit a NOAA weather station broadcast?
A. If the NOAA broadcast is taped and retransmitted later
B. If a general state of communications emergency is declared by the FCC
C. If permission is granted by NOAA for amateur retransmission of the broadcast
D. Under no circumstances

3AA-12.5   Under what circumstances, if any, may an amateur station be used for an activity related to program production or newsgathering for broadcast purposes?
A. The programs or news produced with the assistance of an amateur station must be taped for broadcast at a later time
B. An amateur station may be used for newsgathering and program production only by National Public Radio
C. Under no circumstances
D. Programs or news produced with the assistance of an amateur station must mention the call sign of that station

3AA-13.1   What kinds of one-way communications by amateur stations are not considered broadcasting?
A. All types of one-way communications by amateurs are considered by the FCC as broadcasting
B. Beacon operation, radio-control operation, emergency communications, information bulletins consisting solely of subject matter relating to Amateur Radio, roundtable discussions and code-practice transmissions
C. Only code-practice transmissions conducted simultaneously on all available amateur bands below 30 MHz and conducted for more than 40 hours per week are not considered broadcasting
D. Only actual emergency communications during a declared communications emergency are exempt

3AA-13.2   What is a *one-way radio communication*?
A. A communication in which propagation at the frequency in use supports signal travel in only one direction
B. A communication in which different emissions are used in each direction
C. A communication in which an amateur station transmits to and receives from a station in a radio service other than amateur
D. A transmission to which no on-the-air response is desired or expected

3AA-13.3   What kinds of one-way information bulletins may be transmitted by
            amateur stations?
            A. NOAA weather bulletins
            B. Commuter traffic reports from local radio stations
            C. Regularly scheduled announcements concerning Amateur
               Radio equipment for sale or trade
            D. Bulletins consisting solely of information relating to Amateur
               Radio

3AA-13.4   What types of one-way Amateur Radio communications may be
            transmitted by an amateur station?
            A. Beacon operation, radio control, code practice, retransmission
               of other services
            B. Beacon operation, radio control, transmitting an unmodulated
               carrier, NOAA weather bulletins
            C. Beacon operation, radio control, information bulletins
               consisting solely of information relating to Amateur Radio,
               code practice and emergency communications
            D. Beacon operation, emergency-drill-practice transmissions,
               automatic retransmission of NOAA weather transmissions,
               code practice

3AA-14.1   What types of material compensation, if any, may be involved in
            third-party traffic transmitted by an amateur station?
            A. Payment of an amount agreed upon by the amateur operator
               and the parties involved
            B. Assistance in maintenance of auxiliary station equipment
            C. Donation of amateur equipment to the control operator
            D. No compensation may be accepted

3AA-14.2   What types of business communications, if any, may be transmitted
            by an amateur station on behalf of a third party?
            A. Section 97.57 specifically prohibits business communications
               in the Amateur Service
            B. Business communications involving the sale of Amateur Radio
               equipment
            C. Business communications involving an emergency, as defined
               in Part 97
            D. Business communications aiding a broadcast station

3AA-14.3   Does the FCC allow third-party messages when communicating with
            Amateur Radio operators in a foreign country?
            A. Third-party messages with a foreign country are only allowed
               on behalf of other amateurs.
            B. Yes, provided the third-party message involves the immediate
               family of one of the communicating amateurs
            C. Under no circumstances may US amateurs exchange
               third-party messages with an amateur in a foreign country
            D. Yes, when communicating with a person in a country with
               which the US shares a third-party agreement

3AA-15.1   Under what circumstances, if any, may a third party participate in
            radio communications from an amateur station?
            A. A control operator must be present and continuously monitor
               and supervise the radio communication to ensure compliance
               with the rules. Also, contacts may only be made with amateurs
               in the US and countries with which the US has a third-party
               traffic agreement
            B. A control operator must be present and continuously monitor
               and supervise the radio communication to ensure compliance
               with the rules only if contacts are made with amateurs in
               countries with which the US has no third-party traffic
               agreement
            C. A control operator must be present and continuously monitor
               and supervise the radio communication to ensure compliance
               with the rules. In addition, the control operator must key
               the transmitter and make the station identification.
            D. A control operator must be present and continuously monitor
               and supervise the radio communication to ensure compliance
               with the rules. Also, if contacts are made on frequencies below
               30 MHz, the control operator must transmit the call signs of
               both stations

3AA-15.2   Where must the control operator be situated when a third party is
            participating in radio communications from an amateur station?
            A. If a radio remote control is used, the control operator may be
               physically separated from the control point, when provisions
               are incorporated to shut off the transmitter by remote control
            B. If the control operator supervises the third party until he or she
               is satisfied of the competence of the third party, the control
               operator may leave the control point
            C. The control operator must stay at the control point for the
               entire time the third party is participating
            D. If the third party holds a valid radiotelegraph license issued by
               the FCC, no supervision is necessary

3AA-15.3   What must the control operator do while a third party is participating
            in radio communications?
            A. If the third party holds a valid commercial radiotelegraph
               license, no supervision is necessary
            B. The control operator must tune up and down 5 kHz from the
               transmitting frequency on another receiver, to ensure that no
               interference is taking place
            C. If a radio control link is available, the control operator may
               leave the room
            D. The control operator must continuously monitor and supervise
               the radio communication to ensure compliance with the rules

3AA-15.4   Under what circumstances, if any, may a third party assume the
            duties of the control operator of an amateur station?
            A. If the third party holds a valid commercial radiotelegraph
               license, he or she may act as control operator
            B. Under no circumstances may a third party assume the duties
               of control operator
            C. During Field Day, the third party may act as control operator
            D. An Amateur Extra class licensee may designate a third party
               as control operator, if the station is operated above 450 MHz

3AA-16.1   Under what circumstances, if any, may an amateur station transmit radio communications containing obscene words?
   A. Obscene words are permitted when they do not cause interference to any other radio communication or signal
   B. Obscene words are prohibited in Amateur Radio transmissions
   C. Obscene words are permitted when they are not retransmitted through repeater or auxiliary stations
   D. Obscene words are permitted, but there is an unwritten rule among amateurs that they should not be used on the air

3AA-16.2   Under what circumstances, if any, may an amateur station transmit radio communications containing indecent words?
   A. Indecent words are permitted when they do not cause interference to any other radio communication or signal
   B. Indecent words are permitted when they are not retransmitted through repeater or auxiliary stations
   C. Indecent words are permitted, but there is an unwritten rule among amateurs that they should not be used on the air
   D. Indecent words are prohibited in Amateur Radio transmissions

3AA-16.3   Under what circumstances, if any, may an amateur station transmit radio communications containing profane words?
   A. Profane words are permitted when they are not retransmitted through repeater or auxiliary stations
   B. Profane words are permitted, but there is an unwritten rule among amateurs that they should not be used on the air
   C. Profane words are prohibited in Amateur Radio transmissions
   D. Profane words are permitted when they do not cause interference to any other radio communication or signal

3AA-17.1   What classes of Amateur Radio operator license are eligible for earth operation in the Amateur-Satellite Service?
   A. Novice, Technician, General, Advanced and Amateur Extra class
   B. Technician, General, Advanced and Amateur Extra class
   C. General, Advanced and Amateur Extra class
   D. Amateur Extra class only

SUBELEMENT 3AB—Operating Procedures (3 Exam Questions)

3AB-1.1   What is the meaning of: "Your report is five seven..."?
                 A. Your signal is perfectly readable and moderately strong
                 B. Your signal is perfectly readable, but weak
                 C. Your signal is readable with considerable difficulty
                 D. Your signal is perfectly readable with near pure tone

3AB-1.2   What is the meaning of: "Your report is three three..."?
                 A. The contact is serial number thirty-three
                 B. The station is located at latitude 33 degrees
                 C. Your signal is readable with considerable difficulty and weak in strength
                 D. Your signal is unreadable, very weak in strength

3AB-1.3   What is the meaning of: "Your report is five nine plus 20 dB..."?
                 A. Your signal strength has increased by a factor of 100
                 B. Repeat your transmission on a frequency 20 kHz higher
                 C. The bandwidth of your signal is 20 decibels above linearity
                 D. A relative signal-strength meter reading is 20 decibels greater than strength 9

3AB-2-1.1   How should a QSO be initiated through a station in repeater operation?
                 A. Say "breaker, breaker 79"
                 B. Call the desired station and then identify your own station
                 C. Call "CQ" three times and identify three times
                 D. Wait for a "CQ" to be called and then answer it

3AB-2-1.2   Why should users of a station in repeater operation pause briefly between transmissions?
                 A. To check the SWR of the repeater
                 B. To reach for pencil and paper for third party traffic
                 C. To listen for any hams wanting to break in
                 D. To dial up the repeater's autopatch

3AB-2-1.3   Why should users of a station in repeater operation keep their transmissions short and thoughtful?
                 A. A long transmission may prevent someone with an emergency from using the repeater
                 B. To see if the receiving station operator is still awake
                 C. To give any non-hams that are listening a chance to respond
                 D. To keep long-distance charges down

3AB-2-1.4   What is the proper procedure to break into an on-going QSO through a station in repeater operation?
                 A. Wait for the end of a transmission and start calling
                 B. Shout, "break, break!" to show that you're eager to join the conversation
                 C. Turn on your 100-watt amplifier and override whoever is talking
                 D. Send your call sign during a break between transmissions

3AB-2-1.5   What is the purpose of repeater operation?
                 A. To cut your power bill by using someone's higher power system
                 B. To enable mobile and low-power stations to extend their usable range
                 C. To reduce your telephone bill
                 D. To call the ham radio distributor 50 miles away

3AB-2-1.6 What is meant by "making the repeater time out"?
- A. The repeater's battery supply has run out
- B. The repeater's transmission time limit has expired during a single transmission
- C. The warranty on the repeater duplexer has expired
- D. The repeater is in need of repairs

3AB-2-1.7 During commuting rush hours, which types of operation should relinquish the use of the repeater?
- A. Mobile operators
- B. Low-power stations
- C. Highway traffic information nets
- D. Third-party traffic nets

3AB-2-2.1 Why should simplex be used where possible instead of using a station in repeater operation?
- A. Farther distances can be reached
- B. To avoid long distance toll charges
- C. To avoid tying up the repeater unnecessarily
- D. To permit the testing of the effectiveness of your antenna

3AB-2-2.2 When a frequency conflict arises between a simplex operation and a repeater operation, why does good amateur practice call for the simplex operation to move to another frequency?
- A. The repeater's output power can be turned up to ruin the front end of the station in simplex operation
- B. There are more repeaters than simplex operators
- C. Changing the repeater's frequency is not practical
- D. Changing a repeater frequency requires the authorization of the Federal Communications Commission

3AB-2-3.1 What is the usual input/output frequency separation for stations in repeater operation in the 2-meter band?
- A. 1 MHz
- B. 1.6 MHz
- C. 170 Hz
- D. 0.6 MHz

3AB-2-3.2 What is the usual input/output frequency separation for stations in repeater operation in the 70-centimeter band?
- A. 1.6 MHz
- B. 5 MHz
- C. 600 kHz
- D. 5 kHz

3AB-2-3.3 What is the usual input/output frequency separation for a 6-meter station in repeater operation?
- A. 1 MHz
- B. 600 kHz
- C. 1.6 MHz
- D. 20 kHz

3AB-2-3.4 What is the usual input/output frequency separation for a 1.25-meter station in repeater operation?
- A. 1000 kHz
- B. 600 kHz
- C. 1600 kHz
- D. 1.6 GHz

3AB-2-4.1   What is a *repeater frequency coordinator*?
             A. Someone who coordinates the assembly of a repeater station
             B. Someone who provides advice on what kind of system to buy
             C. The club's repeater trustee
             D. A person or group that recommends frequency pairs for
                repeater usage

3AB-3.1   Why should local Amateur Radio communications be conducted on
             VHF and UHF frequencies?
             A. To minimize interference on HF bands capable of long-
                distance sky-wave communication
             B. Because greater output power is permitted on VHF and UHF
             C. Because HF transmissions are not propagated locally
             D. Because absorption is greater at VHF and UHF frequencies

3AB-3.2   How can on-the-air transmissions be minimized during a lengthy
             transmitter testing or loading up procedure?
             A. Choose an unoccupied frequency
             B. Use a dummy antenna
             C. Use a non-resonant antenna
             D. Use a resonant antenna that requires no loading up procedure

3AB-3.3   What is the proper Q signal to use to determine whether a
             frequency is in use before making a transmission?
             A. QRV?
             B. QRU?
             C. QRL?
             D. QRZ?

3AB-4.1   What is the proper distress calling procedure when using
             telephony?
             A. Transmit MAYDAY
             B. Transmit QRRR
             C. Transmit QRZ
             D. Transmit SOS

3AB-4.2   What is the proper distress calling procedure when using
             telegraphy?
             A. Transmit MAYDAY
             B. Transmit QRRR
             C. Transmit QRZ
             D. Transmit SOS

3AB-5-1.1   What is one requirement you must meet before you can participate
             in RACES drills?
             A. You must be registered with ARRL
             B. You must be registered with a local racing organization
             C. You must be registered with the responsible civil defense
                organization
             D. You need not register with anyone to operate RACES

3AB-5-1.2   What is the maximum amount of time allowed per week for RACES
             drills?
             A. Eight hours
             B. One hour
             C. As many hours as you want
             D. Six hours, but not more than one hour per day

3AB-5-2.1   How must you identify messages sent during a RACES drill?
A. As emergency messages
B. As amateur traffic
C. As official government messages
D. As drill or test messages

3AB-6-1.1   What is the term used to describe first-response communications in an emergency situation?
A. Tactical communications
B. Emergency communications
C. Formal message traffic
D. National Traffic System messages

3AB-6-1.2   What is one reason for using tactical call signs such as "command post" or "weather center" during an emergency?
A. They keep the general public informed about what is going on
B. They promote efficiency and coordination in public-service communications activities
C. They are required by the FCC
D. They promote goodwill among amateurs

3AB-6-2.1   What is the term used to describe messages sent into or out of a disaster area that pertain to a person's well being?
A. Emergency traffic
B. Tactical traffic
C. Formal message traffic
D. Health and welfare traffic

3AB-6-3.1   Why is it important to provide a means of operating your Amateur Radio station separate from the commercial ac power lines?
A. So that you can take your station mobile
B. So that you can provide communications in an emergency
C. So that you can operate field day
D. So that you will comply with Subpart 97.169 of the FCC Rules

3AB-6-3.2   Which type of antenna would be a good choice as part of a portable HF Amateur Radio station that could be set up in case of a communications emergency?
A. A three-element quad
B. A three-element Yagi
C. A dipole
D. A parabolic dish

3AC-1-1.1   What is the *ionosphere*?
          A. That part of the upper atmosphere where enough ions and free electrons exist to affect radio-wave propagation
          B. The boundary between two air masses of different temperature and humidity, along which radio waves can travel
          C. The ball that goes on the top of a mobile whip antenna
          D. That part of the atmosphere where weather takes place

3AC-1-1.2   What is the region of the outer atmosphere that makes long-distance radio communications possible as a result of bending of radio waves?
          A. Troposphere
          B. Stratosphere
          C. Magnetosphere
          D. Ionosphere

3AC-1-1.3   What type of solar radiation is most responsible for ionization in the outer atmosphere?
          A. Thermal
          B. Ionized particle
          C. Ultraviolet
          D. Microwave

3AC-1-2.1   Which ionospheric layer limits daytime radio communications in the 80-meter band to short distances?
          A. D layer
          B. F1 layer
          C. E layer
          D. F2 layer

3AC-1-2.2   What is the lowest ionospheric layer?
          A. The A layer
          B. The D layer
          C. The E layer
          D. The F layer

3AC-1-3.1   What is the lowest region of the ionosphere that is useful for long-distance radio wave propagation?
          A. The D layer
          B. The E layer
          C. The F1 layer
          D. The F2 layer

3AC-1-4.1   Which layer of the ionosphere is mainly responsible for long-distance sky-wave radio communications?
          A. D layer
          B. E layer
          C. F1 layer
          D. F2 layer

3AC-1-4.2   What are the two distinct sub-layers of the F layer of the ionosphere during the daytime?
          A. Troposphere and stratosphere
          B. F1 and F2
          C. Electrostatic and electromagnetic
          D. D and E

3AC-1-4.3 Which two daytime ionospheric layers combine into one layer at night?
A. E and F1
B. D and E
C. F1 and F2
D. E1 and E2

3AC-2.1 Which layer of the ionosphere is most responsible for absorption of radio signals during daylight hours?
A. The E layer
B. The F1 layer
C. The F2 layer
D. The D layer

3AC-2.2 When is ionospheric absorption most pronounced?
A. When tropospheric ducting occurs
B. When radio waves enter the D layer at low angles
C. When radio waves travel to the F layer
D. When a temperature inversion occurs

3AC-2.3 During daylight hours, what effect does the D layer of the ionosphere have on 80-meter radio waves?
A. The D layer absorbs the signals
B. The D layer bends the radio waves out into space
C. The D layer refracts the radio waves back to earth
D. The D layer has little or no effect on 80 meter radio wave propagation

3AC-2.4 What causes *ionospheric absorption* of radio waves?
A. A lack of D layer ionization
B. D layer ionization
C. The presence of ionized clouds in the E layer
D. Splitting of the F layer

3AC-3.1 What is usually the condition of the ionosphere just before sunrise?
A. Atmospheric attenuation is at a maximum
B. Ionization is at a maximum
C. The E layer is above the F layer
D. Ionization is at a minimum

3AC-3.2 At what time of day does maximum ionization of the ionosphere occur?
A. Dusk
B. Midnight
C. Midday
D. Dawn

3AC-3.3 Minimum ionization of the ionosphere occurs daily at what time?
A. Shortly before dawn
B. Just after noon
C. Just after dusk
D. Shortly before midnight

3AC-3.4 When is E layer ionization at a maximum?
A. Dawn
B. Midday
C. Dusk
D. Midnight

3AC-4.1   What is the name for the highest radio frequency that will be refracted back to earth?
A. Lowest usable frequency
B. Optimum working frequency
C. Ultra high frequency
D. Critical frequency

3AC-4.2   What causes the *maximum usable frequency* to vary?
A. Variations in the temperature of the air at ionospheric levels
B. Upper-atmospheric wind patterns
C. The amount of ultraviolet and other types of radiation received from the sun
D. Presence of ducting

3AC-4.3   What does the term *maximum usable frequency* refer to?
A. The maximum frequency that allows a radio signal to reach its destination in a single hop
B. The minimum frequency that allows a radio signal to reach its destination in a single hop
C. The maximum frequency that allows a radio signal to be absorbed in the lowest ionospheric layer
D. The minimum frequency that allows a radio signal to be absorbed in the lowest ionospheric layer

3AC-5.1   When two stations are within each other's skip zone on the frequency being used, what mode of propagation would it be desirable to use?
A. Ground-wave propagation
B. Sky-wave propagation
C. Scatter-mode propagation
D. Ionospheric-ducting propagation

3AC-5.2   You are in contact with a distant station and are operating at a frequency close to the maximum usable frequency. If the received signals are weak and somewhat distorted, what type of propagation are you probably experiencing?
A. Tropospheric ducting
B. Line-of-sight propagation
C. Backscatter propagation
D. Waveguide propagation

3AC-6.1   What is the transmission path of a wave that travels directly from the transmitting antenna to the receiving antenna called?
A. Line of sight
B. The sky wave
C. The linear wave
D. The plane wave

3AC-6.2   How are VHF signals within the range of the visible horizon propagated?
A. By sky wave
B. By direct wave
C. By plane wave
D. By geometric wave

3AC-7.1   Ducting occurs in which region of the atmosphere?
A. F2
B. Ionosphere
C. Troposphere
D. Stratosphere

3AC-7.2   What effect does tropospheric bending have on 2-meter radio waves?
A. It increases the distance over which they can be transmitted
B. It decreases the distance over which they can be transmitted
C. It tends to garble 2-meter phone transmissions
D. It reverses the sideband of 2-meter phone transmissions

3AC-7.3   What atmospheric phenomenon causes tropospheric ducting of radio waves?
A. A very low pressure area
B. An aurora to the north
C. Lightning between the transmitting and receiving station
D. A temperature inversion

3AC-7.4   Tropospheric ducting occurs as a result of what phenomenon?
A. A temperature inversion
B. Sun spots
C. An aurora to the north
D. Lightning between the transmitting and receiving station

3AC-7.5   What atmospheric phenomenon causes VHF radio waves to be propagated several hundred miles through stable air masses over oceans?
A. Presence of a maritime polar air mass
B. A widespread temperature inversion
C. An overcast of cirriform clouds
D. Atmospheric pressure of roughly 29 inches of mercury or higher

3AC-7.6   In what frequency range does tropospheric ducting occur most often?
A. LF
B. MF
C. HF
D. VHF

3AD-1-1.1   Where should the green wire in an ac line cord be attached in a power supply?
     A. To the fuse
     B. To the "hot" side of the power switch
     C. To the chassis
     D. To the meter

3AD-1-1.2   Where should the black (or red) wire in a three-wire line cord be attached in a power supply?
     A. To the filter capacitor
     B. To the dc ground
     C. To the chassis
     D. To the fuse

3AD-1-1.3   Where should the white wire in a three-wire line cord be attached in a power supply?
     A. To the side of the transformer's primary winding that has a fuse
     B. To the side of the transformer's primary winding without a fuse
     C. To the black wire
     D. To the rectifier junction

3AD-1-1.4   Why is the retaining screw in one terminal of a light socket made of brass while the other one is silver colored?
     A. To prevent galvanic action
     B. To indicate correct wiring polarity
     C. To better conduct current
     D. To reduce skin effect

3AD-1-2.1   How much electrical current flowing through the human body is usually fatal?
     A. As little as 100 milliamperes may be fatal
     B. Approximately 10 amperes is required to be fatal
     C. More than 20 amperes is needed to kill a human being
     D. No amount of current will harm you. Voltages of over 2000 volts are always fatal, however

3AD-1-2.2   What is the minimum voltage considered to be dangerous to humans?
     A. 30 volts
     B. 100 volts
     C. 1000 volts
     D. 2000 volts

3AD-1-2.3   How much electrical current flowing through the human body is usually painful?
     A. As little as 50 milliamperes may be painful
     B. Approximately 10 amperes is required to be painful
     C. More than 20 amperes is needed to be painful to a human being
     D. No amount of current will be painful. Voltages of over 2000 volts are always painful, however

3AD-1-3.1 Where should the main power-line switch for a high voltage power supply be situated?
A. Inside the cabinet, to interrupt power when the cabinet is opened
B. On the rear panel of the high-voltage supply
C. Where it can be seen and reached easily
D. This supply should not be switch-operated

3AD-2-1.1 How is a voltmeter typically connected to a circuit under test?
A. In series with the circuit
B. In parallel with the circuit
C. In quadrature with the circuit
D. In phase with the circuit

3AD-2-2.1 How can the range of a voltmeter be extended?
A. By adding resistance in series with the circuit under test
B. By adding resistance in parallel with the circuit under test
C. By adding resistance in series with the meter
D. By adding resistance in parallel with the meter

3AD-3-1.1 How is an ammeter typically connected to a circuit under test?
A. In series with the circuit
B. In parallel with the circuit
C. In quadrature with the circuit
D. In phase with the circuit

3AD-3-2.1 How can the range of an ammeter be extended?
A. By adding resistance in series with the circuit under test
B. By adding resistance in parallel with the circuit under test
C. By adding resistance in series with the meter
D. By adding resistance in parallel with the meter

3AD-4.1 What is a *multimeter*?
A. An instrument capable of reading SWR and power
B. An instrument capable of reading resistance, capacitance and inductance
C. An instrument capable of reading resistance and reactance
D. An instrument capable of reading voltage, current and resistance

3AD-5-1.1 Where in the antenna transmission line should a peak-reading wattmeter be attached to determine the transmitter output power?
A. At the transmitter output
B. At the antenna feed point
C. One-half wavelength from the antenna feed point
D. One-quarter wavelength from the transmitter output

3AD-5-1.2 For the most accurate readings of transmitter output power, where should the RF wattmeter be inserted?
A. The wattmeter should be inserted and the output measured one-quarter wavelength from the antenna feed point
B. The wattmeter should be inserted and the output measured one-half wavelength from the antenna feed point
C. The wattmeter should be inserted and the output power measured at the transmitter antenna jack
D. The wattmeter should be inserted and the output power measured at the Transmatch output

3AD-5-1.3   At what line impedance are RF wattmeters usually designed to operate?
A. 25 ohms
B. 50 ohms
C. 100 ohms
D. 300 ohms

3AD-5-1.4   What is a *directional wattmeter*?
A. An instrument that measures forward or reflected power
B. An instrument that measures the directional pattern of an antenna
C. An instrument that measures the energy consumed by the transmitter
D. An instrument that measures thermal heating in a load resistor

3AD-5-2.1   If a directional RF wattmeter indicates 90 watts forward power and 10 watts reflected power, what is the actual transmitter output power?
A. 10 watts
B. 80 watts
C. 90 watts
D. 100 watts

3AD-5-2.2   If a directional RF wattmeter indicates 96 watts forward power and 4 watts reflected power, what is the actual transmitter output power?
A. 80 watts
B. 88 watts
C. 92 watts
D. 100 watts

3AD-6.1   What is a *marker generator*?
A. A high-stability oscillator that generates a series of reference signals at known frequency intervals
B. A low-stability oscillator that "sweeps" through a band of frequencies
C. An oscillator often used in aircraft to determine the craft's location relative to the inner and outer markers at airports
D. A high-stability oscillator whose output frequency and amplitude can be varied over a wide range

3AD-6.2   What type of circuit is used to inject a frequency calibration signal into a communications receiver?
A. A product detector
B. A receiver incremental tuning circuit
C. A balanced modulator
D. A crystal calibrator

3AD-6.3   How is a *marker generator* used?
A. To calibrate the tuning dial on a receiver
B. To calibrate the volume control on a receiver
C. To test the amplitude linearity of an SSB transmitter
D. To test the frequency deviation of an FM transmitter

3AD-7.1   What piece of test equipment produces a stable, low-level signal that can be set to a specific frequency?
A. A wavemeter
B. A reflectometer
C. A signal generator
D. A balanced modulator

3AD-7.2 What is an *RF signal generator* commonly used for?
  A. Measuring RF signal amplitude
  B. Aligning receiver tuned circuits
  C. Adjusting the transmitter impedance-matching network
  D. Measuring transmission line impedance

3AD-8-1.1 What is a *reflectometer*?
  A. An instrument used to measure signals reflected from the ionosphere
  B. An instrument used to measure radiation resistance
  C. An instrument used to measure transmission-line impedance
  D. An instrument used to measure standing wave ratio

3AD-8-1.2 What is the device that can indicate an impedance mismatch in an antenna system?
  A. A field-strength meter
  B. A set of lecher wires
  C. A wavemeter
  D. A reflectometer

3AD-8-2.1 For best accuracy when adjusting the impedance match between an antenna and feed line, where should the match-indicating device be inserted?
  A. At the antenna feed point
  B. At the transmitter
  C. At the midpoint of the feed line
  D. Anywhere along the feed line

3AD-8-2.2 Where should a reflectometer be inserted into a long antenna transmission line in order to obtain the most valid standing wave ratio indication?
  A. At any quarter-wavelength interval along the transmission line
  B. At the receiver end
  C. At the antenna end
  D. At any even half-wavelength interval along the transmission line

3AD-9.1 When adjusting a transmitter filter circuit, what device is connected to the transmitter output?
  A. A multimeter
  B. A set of Litz wires
  C. A receiver
  D. A dummy antenna

3AD-9.2 What is a *dummy antenna*?
  A. An isotropic radiator
  B. A nonradiating load for a transmitter
  C. An antenna used as a reference for gain measurements
  D. The image of an antenna, located below ground

3AD-9.3 Of what materials may a dummy antenna be made?
  A. A wire-wound resistor
  B. A diode and resistor combination
  C. A noninductive resistor
  D. A coil and capacitor combination

3AD-9.4   What station accessory is used in place of an antenna during
            transmitter tests so that no signal is radiated?
            A. A Transmatch
            B. A dummy antenna
            C. A low-pass filter
            D. A decoupling resistor

3AD-9.5   What is the purpose of a *dummy load*?
            A. To allow off-the-air transmitter testing
            B. To reduce output power for QRP operation
            C. To give comparative signal reports
            D. To allow Transmatch tuning without causing interference

3AD-9.6   How many watts should a dummy load for use with a 100-watt
            emission J3E transmitter with 50 ohm output be able to dissipate?
            A. A minimum of 100 watts continuous
            B. A minimum of 141 watts continuous
            C. A minimum of 175 watts continuous
            D. A minimum of 200 watts continuous

3AD-10.1   What is an *S-meter*?
            A. A meter used to measure sideband suppression
            B. A meter used to measure spurious emissions from a
               transmitter
            C. A meter used to measure relative signal strength in a receiver
            D. A meter used to measure solar flux

3AD-10.2   A meter that is used to measure relative signal strength in a
            receiver is known as what?
            A. An S-meter
            B. An RST-meter
            C. A signal deviation meter
            D. An SSB meter

3AD-11-1.1 Large amounts of RF energy may cause damage to body tissue,
            depending on the wavelength of the signal, the energy density of
            the RF field, and other factors. How does RF energy effect body
            tissue?
            A. It causes radiation poisioning
            B. It heats the tissue
            C. It cools the tissue
            D. It produces genetic changes in the tissue

3AD-11-1.2 Which body organ is most susceptible to damage from the heating
            effects of radio frequency radiation?
            A. Eyes
            B. Hands
            C. Heart
            D. Liver

3AD-11-2.1 Scientists have devoted a great deal of effort to determine safe RF
            exposure limits. What organization has established an RF protection
            guide?
            A. The Institute of Electrical and Electronics Engineers
            B. The American Radio Relay League
            C. The Environmental Protection Agency
            D. The American National Standards Institute

3AD-11-2.2   What is the purpose of the ANSI RF protection guide?
               A. It protects you from unscrupulous radio dealers
               B. It sets RF exposure limits under certain circumstances
               C. It sets transmitter power limits
               D. It sets antenna height requirements

3AD-11-2.3   The American National Standards Institute RF protection guide sets RF exposure limits under certain circumstances. In what frequency range is the maximum exposure level the most stringent (lowest)?
               A. 3 to 30 MHz
               B. 30 to 300 MHz
               C. 300 to 3000 MHz
               D. Above 1.5 GHz

3AD-11-2.4   The American National Standards Institute RF protection guide sets RF exposure limits under certain circumstances. Why is the maximum exposure level the most stringent (lowest) in the ranges between 30 MHz and 300 MHz?
               A. There are fewer transmitters operating in this frequency range
               B. There are more transmitters operating in this frequency range
               C. Most transmissions in this frequency range are for an extended time
               D. Human body lengths are close to whole-body resonance in that range

3AD-11-2.5   The American National Standards Institute RF protection guide sets RF exposure limits under certain circumstances. What is the maximum safe power output to the antenna terminal of a hand-held VHF or UHF radio, as set by this RF protection guide?
               A. 125 milliwatts
               B. 7 watts
               C. 10 watts
               D. 25 watts

3AD-11-3.1   After you make internal tuning adjustments to your VHF power amplifier, what should you do before you turn the amplifier on?
               A. Remove all amplifier shielding to ensure maximum cooling
               B. Connect a noise bridge to eliminate any interference
               C. Be certain all amplifier shielding is fastened in place
               D. Be certain no antenna is attached so that you will not cause any interference

SUBELEMENT 3AE—Electrical Principles (2 Exam Questions)

3AE-1-1.1   What is meant by the term *resistance*?
    A. The opposition to the flow of current in an electric circuit containing inductance
    B. The opposition to the flow of current in an electric circuit containing capacitance
    C. The opposition to the flow of current in an electric circuit containing reactance
    D. The opposition to the flow of current in an electric circuit that does not contain reactance

3AE-1-2.1   What is an *ohm*?
    A. The basic unit of resistance
    B. The basic unit of capacitance
    C. The basic unit of inductance
    D. The basic unit of admittance

3AE-1-2.2   What is the unit measurement of resistance?
    A. Volt
    B. Ampere
    C. Joule
    D. Ohm

3AE-1-3.1   Two equal-value resistors are connected in series. How does the total resistance of this combination compare with the value of either resistor by itself?
    A. The total resistance is half the value of either resistor
    B. The total resistance is twice the value of either resistor
    C. The total resistance is the same as the value of either resistor
    D. The total resistance is the square of the value of either resistor

3AE-1-3.2   How does the total resistance of a string of series-connected resistors compare to the values of the individual resistors?
    A. The total resistance is the square of the sum of all the individual resistor values
    B. The total resistance is the square root of the sum of the individual resistor values
    C. The total resistance is the sum of the squares of the individual resistor values
    D. The total resistance is the sum of all the individual resistance values

3AE-1-4.1   Two equal-value resistors are connected in parallel. How does the total resistance of this combination compare with the value of either resistor by itself?
    A. The total resistance is twice the value of either resistor
    B. The total resistance is half the value of either resistor
    C. The total resistance is the square of the value of either resistor
    D. The total resistance is the same as the value of either resistor

3AE-1-4.2   How does the total resistance of a string of parallel-connected resistors compare to the values of the individual resistors?

    A. The total resistance is the square of the sum of the resistor values

    B. The total resistance is more than the highest-value resistor in the combination

    C. The total resistance is less than the smallest-value resistor in the combination

    D. The total resistance is the same as the highest-value resistor in the combination

3AE-2.1   What is *Ohm's Law*?

    A. A mathematical relationship between resistance, voltage and power in a circuit

    B. A mathematical relationship between current, resistance and power in a circuit

    C. A mathematical relationship between current, voltage and power in a circuit

    D. A mathematical relationship between resistance, current and applied voltage in a circuit

3AE-2.2   How is the current in a dc circuit calculated when the voltage and resistance are known?

    A. $I = E / R$

    B. $P = I \times E$

    C. $I = R \times E$

    D. $I = E \times R$

3AE-2.3   What is the input resistance of a load when a 12-volt battery supplies 0.25 amperes to it?

    A. 0.02 ohms

    B. 3 ohms

    C. 48 ohms

    D. 480 ohms

3AE-2.4   The product of the current and what force gives the electrical power in a circuit?

    A. Magnetomotive force

    B. Centripetal force

    C. Electrochemical force

    D. Electromotive force

3AE-2.5   What is the input resistance of a load when a 12-volt battery supplies 0.15 amperes to it?

    A. 8 ohms

    B. 80 ohms

    C. 100 ohms

    D. 800 ohms

3AE-2.6   When 120 volts is measured across a 4700-ohm resistor, approximately how much current is flowing through it?

    A. 39 amperes

    B. 3.9 amperes

    C. 0.26 ampere

    D. 0.026 ampere

3AE-2.7 When 120 volts is measured across a 47000-ohm resistor, approximately how much current is flowing through it?
A. 392 A
B. 39.2 A
C. 26 mA
D. 2.6 mA

3AE-2.8 When 12 volts is measured across a 4700-ohm resistor, approximately how much current is flowing through it?
A. 2.6 mA
B. 26 mA
C. 39.2 A
D. 392 A

3AE-2.9 When 12 volts is measured across a 47000-ohm resistor, approximately how much current is flowing through it?
A. 255 μA
B. 255 mA
C. 3917 mA
D. 3917 A

3AE-3-1.1 What is the term used to describe the ability of a component to store energy in a magnetic field?
A. Admittance
B. Capacitance
C. Inductance
D. Resistance

3AE-3-2.1 What is the basic unit of inductance?
A. Coulomb
B. Farad
C. Henry
D. Ohm

3AE-3-2.2 What is a *henry*?
A. The basic unit of admittance
B. The basic unit of capacitance
C. The basic unit of inductance
D. The basic unit of resistance

3AE-3-2.3 What is a *microhenry*?
A. A basic unit of inductance equal to 10^{-12} henrys
B. A basic unit of inductance equal to 10^{-6} henrys
C. A basic unit of inductance equal to 10^{-3} henrys
D. A basic unit of inductance equal to 10^6 henrys

3AE-3-2.4 What is a *millihenry*?
A. A basic unit of inductance equal to 10^{-12} henrys
B. A basic unit of inductance equal to 10^{-6} henrys
C. A basic unit of inductance equal to 10^{-3} henrys
D. A basic unit of inductance equal to 10^6 henrys

3AE-3-3.1 Two equal-value inductors are connected in series. How does the total inductance of this combination compare with the value of either inductor by itself?
A. The total inductance is half the value of either inductor
B. The total inductance is twice the value of either inductor
C. The total inductance is equal to the value of either inductor
D. No comparison can be made without knowing the exact inductances

3AE-3-3.2 How does the total inductance of a string of series-connected inductors compare to the values of the individual inductors?
- A. The total inductance is equal to the average of all the individual inductances
- B. The total inductance is equal to less than the value of the smallest inductance
- C. The total inductance is equal to the sum of all the individual inductances
- D. No comparison can be made without knowing the exact inductances

3AE-3-4.1 Two equal-value inductors are connected in parallel. How does the total inductance of this combination compare with the value of either inductor by itself?
- A. The total inductance is half the value of either inductor
- B. The total inductance is twice the value of either inductor
- C. The total inductance is equal to the square of either inductance
- D. No comparison can be made without knowing the exact inductances

3AE-3-4.2 How does the total inductance of a string of parallel-connected inductors compare to the values of the individual inductors?
- A. The total inductance is equal to the sum of the inductances in the combination
- B. The total inductance is less than the smallest inductance value in the combination
- C. The total inductance is equal to the average of the inductances in the combination
- D. No comparison can be made without knowing the exact inductances

3AE-4-1.1 What is the term used to describe the ability of a component to store energy in an electric field?
- A. Capacitance
- B. Inductance
- C. Resistance
- D. Tolerance

3AE-4-2.1 What is the basic unit of capacitance?
- A. Farad
- B. Ohm
- C. Volt
- D. Ampere

3AE-4-2.2 What is a *microfarad*?
- A. A basic unit of capacitance equal to 10^{-12} farads
- B. A basic unit of capacitance equal to 10^{-6} farads
- C. A basic unit of capacitance equal to 10^{-2} farads
- D. A basic unit of capacitance equal to 10^{6} farads

3AE-4-2.3 What is a *picofarad*?
- A. A basic unit of capacitance equal to 10^{-12} farads
- B. A basic unit of capacitance equal to 10^{-6} farads
- C. A basic unit of capacitance equal to 10^{-2} farads
- D. A basic unit of capacitance equal to 10^{6} farads

3AE-4-2.4   What is a *farad*?
        A. The basic unit of resistance
        B. The basic unit of capacitance
        C. The basic unit of inductance
        D. The basic unit of admittance

3AE-4-3.1   Two equal-value capacitors are connected in series. How does the total capacitance of this combination compare with the value of either capacitor by itself?
        A. The total capacitance is twice the value of either capacitor
        B. The total capacitance is equal to the value of either capacitor
        C. The total capacitance is half the value of either capacitor
        D. No comparison can be made without knowing the exact capacitances

3AE-4-3.2   How does the total capacitance of a string of series-connected capacitors compare to the values of the individual capacitors?
        A. The total capacitance is equal to the sum of the capacitances in the combination
        B. The total capacitance is less than the smallest value of capacitance in the combination
        C. The total capacitance is equal to the average of the capacitances in the combination
        D. No comparison can be made without knowing the exact capacitances

3AE-4-4.1   Two equal-value capacitors are connected in parallel. How does the total capacitance of this combination compare with the value of either capacitor by itself?
        A. The total capacitance is twice the value of either capacitor
        B. The total capacitance is half the value of either capacitor
        C. The total capacitance is equal to the value of either capacitor
        D. No comparison can be made without knowing the exact capacitances

3AE-4-4.2   How does the total capacitance of a string of parallel-connected capacitors compare to the values of the individual capacitors?
        A. The total capacitance is equal to the sum of the capacitances in the combination
        B. The total capacitance is less than the smallest value of capacitance in the combination
        C. The total capacitance is equal to the average of the capacitances in the combination
        D. No comparison can be made without knowing the exact capacitances

SUBELEMENT 3AF—Circuit Components (2 Exam Questions)

3AF-1-1.1   What are the four common types of resistor construction?
A. Carbon-film, metal-film, micro-film and wire-film
B. Carbon-composition, carbon-film, metal-film and wire-wound
C. Carbon-composition, carbon-film, electrolytic and metal-film
D. Carbon-film, ferrite, carbon-composition and metal-film

3AF-1-2.1   What is the primary function of a resistor?
A. To store an electric charge
B. To store a magnetic field
C. To match a high-impedance source to a low-impedance load
D. To limit the current in an electric circuit

3AF-1-2.2   What is a *variable resistor*?
A. A resistor that changes value when an ac voltage is applied to it
B. A device that can transform a variable voltage into a constant voltage
C. A resistor with a slide or contact that makes the resistance adjustable
D. A resistor that changes value when it is heated

3AF-1-3.1   What do the first three color bands on a resistor indicate?
A. The value of the resistor in ohms
B. The resistance tolerance in percent
C. The power rating in watts
D. The value of the resistor in henrys

3AF-1-3.2   How can a carbon resistor's electrical tolerance rating be found?
A. By using a wavemeter
B. By using the resistor's color code
C. By using Thevenin's theorem for resistors
D. By using the Baudot code

3AF-1-3.3   What does the fourth color band on a resistor indicate?
A. The value of the resistor in ohms
B. The resistance tolerance in percent
C. The power rating in watts
D. The resistor composition

3AF-1-3.4   When the color bands on a group of resistors indicate that they all have the same resistance, what further information about each resistor is needed in order to select those that have nearly equal value?
A. The working voltage rating of each resistor
B. The composition of each resistor
C. The tolerance of each resistor
D. The current rating of each resistor

3AF-1-4.1   Why do resistors generate heat?
A. They convert electrical energy to heat energy
B. They exhibit reactance
C. Because of skin effect
D. To produce thermionic emission

3AF-1-4.2 Why would a large size resistor be substituted for a smaller one of the same resistance?
- A. To obtain better response
- B. To obtain a higher current gain
- C. To increase power dissipation capability
- D. To produce a greater parallel impedance

3AF-1-5.1 What is the symbol used to represent a fixed resistor on schematic diagrams?

A.   B.

C.   D.

3AF-1-5.2 What is the symbol used to represent a variable resistor on schematic diagrams?

A.   B.

C.   D.

3AF-2-1.1 What is an inductor *core*?
- A. The point at which an inductor is tapped to produce resonance
- B. A tight coil of wire used in a transformer
- C. An insulating material placed between the plates of an inductor
- D. The central portion of a coil; may be made from air, iron, brass or other material

3AF-2-1.2 What are the component parts of a coil?
- A. The wire in the winding and the core material
- B. Two conductive plates and an insulating material
- C. Two or more layers of silicon material
- D. A donut-shaped iron core and a layer of insulating tape

3AF-2-1.3 Describe an *inductor*.
- A. A semiconductor in a conducting shield
- B. Two parallel conducting plates
- C. A straight wire conductor mounted inside a Faraday shield
- D. A coil of conducting wire

3AF-2-1.4 For radio frequency power applications, which type of inductor has the least amount of loss?
- A. Magnetic wire
- B. Iron core
- C. Air core
- D. Slug tuned

3AF-2-2.1 What is an *inductor*?
A. An electronic component that stores energy in an electric field
B. An electronic component that converts a high voltage to a lower voltage
C. An electronic component that opposes dc while allowing ac to pass
D. An electronic component that stores energy in a magnetic field

3AF-2-2.2 What are the electrical properties of an inductor?
A. An inductor stores a charge electrostatically and opposes a change in voltage
B. An inductor stores a charge electrochemically and opposes a change in current
C. An inductor stores a charge electromagnetically and opposes a change in current
D. An inductor stores a charge electromechanically and opposes a change in voltage

3AF-2-3.1 What factors determine the amount of inductance in a coil?
A. The type of material used in the core, the diameter of the core and whether the coil is mounted horizontally or vertically
B. The diameter of the core, the number of turns of wire used to wind the coil and the type of metal used in the wire
C. The type of material used in the core, the number of turns used to wind the core and the frequency of the current through the coil
D. The type of material used in the core, the diameter of the core, the length of the coil and the number of turns of wire used to wind the coil

3AF-2-3.2 What can be done to raise the inductance of a 5-microhenry air-core coil to a 5-millihenry coil with the same physical dimensions?
A. The coil can be wound on a non-conducting tube
B. The coil can be wound on an iron core
C. Both ends of the coil can be brought around to form the shape of a donut, or toroid
D. The coil can be made of a heavier-gauge wire

3AF-2-3.3 As an iron core is inserted in a coil, what happens to the inductance?
A. It increases
B. It decreases
C. It stays the same
D. It becomes voltage-dependent

3AF-2-3.4 As a brass core is inserted in a coil, what happens to the inductance?
A. It increases
B. It decreases
C. It stays the same
D. It becomes voltage-dependent

3AF-2-4.1 What is the symbol used to represent an adjustable inductor on schematic diagrams?

A.

B. ⫯⫯

C.

D. ⟋⋁⋁⟍

3AF-2-4.2 What is the symbol used to represent an iron-core inductor on schematic diagrams?

A.

B. ⟋⟍⟍⟍

C.

D.

3AF-2-4.3 What is the symbol used to represent an inductor wound over a toroidal core on schematic diagrams?

A.

B.

C.

D.

3AF-3-1.1 What is a capacitor *dielectric*?
A. The insulating material used for the plates
B. The conducting material used between the plates
C. The ferrite material that the plates are mounted on
D. The insulating material between the plates

3AF-3-1.2 What are the component parts of a capacitor?
A. Two or more conductive plates with an insulating material between them
B. The wire used in the winding and the core material
C. Two or more layers of silicon material
D. Two insulating plates with a conductive material between them

3AF-3-1.3 What is an *electrolytic capacitor*?
A. A capacitor whose plates are formed on a thin ceramic layer
B. A capacitor whose plates are separated by a thin strip of mica insulation
C. A capacitor whose dielectric is formed on one set of plates through electrochemical action
D. A capacitor whose value varies with applied voltage

3AF-3-1.4 What is a *paper capacitor*?
A. A capacitor whose plates are formed on a thin ceramic layer
B. A capacitor whose plates are separated by a thin strip of mica insulation
C. A capacitor whose plates are separated by a layer of paper
D. A capacitor whose dielectric is formed on one set of plates through electrochemical action

3AF-3-2.1 What is a *capacitor*?
A. An electronic component that stores energy in a magnetic field
B. An electronic component that stores energy in an electric field
C. An electronic component that converts a high voltage to a lower voltage
D. An electronic component that converts power into heat

3AF-3-2.2 What are the electrical properties of a capacitor?
A. A capacitor stores a charge electrochemically and opposes a change in current
B. A capacitor stores a charge electromagnetically and opposes a change in current
C. A capacitor stores a charge electromechanically and opposes a change in voltage
D. A capacitor stores a charge electrostatically and opposes a change in voltage

3AF-3-2.3 What factors must be considered when selecting a capacitor for a circuit?
A. Type of capacitor, capacitance and voltage rating
B. Type of capacitor, capacitance and the kilowatt-hour rating
C. The amount of capacitance, the temperature coefficient and the KVA rating
D. The type of capacitor, the microscopy coefficient and the temperature coefficient

3AF-3-2.4 How are the characteristics of a capacitor usually specified?
A. In volts and amperes
B. In microfarads and volts
C. In ohms and watts
D. In millihenrys and amperes

3AF-3-3.1 What factors determine the amount of capacitance in a capacitor?
A. The dielectric constant of the material between the plates, the area of one side of one plate, the separation between the plates and the number of plates
B. The dielectric constant of the material between the plates, the number of plates and the diameter of the leads connected to the plates
C. The number of plates, the spacing between the plates and whether the dielectric material is N type or P type
D. The dielectric constant of the material between the plates, the surface area of one side of one plate, the number of plates and the type of material used for the protective coating

3AF-3-3.2   As the plate area of a capacitor is increased, what happens to its
             capacitance?
             A. Decreases
             B. Increases
             C. Stays the same
             D. Becomes voltage dependent

3AF-3-3.3   As the plate spacing of a capacitor is increased, what happens to
             its capacitance?
             A. Increases
             B. Stays the same
             C. Becomes voltage dependent
             D. Decreases

3AF-3-4.1   What is the symbol used to represent an electrolytic capacitor on
             schematic diagrams?

             A.   B.

             C.   D.

3AF-3-4.2   What is the symbol used to represent a variable capacitor on
             schematic diagrams?

             A.   B.

             C.   D.

3AG-1-1.1   Which frequencies are attenuated by a low-pass filter?
    A. Those above its cut-off frequency
    B. Those within its cut-off frequency
    C. Those within 50 kHz on either side of its cut-off frequency
    D. Those below its cut-off frequency

3AG-1-1.2   What circuit passes electrical energy below a certain frequency and blocks electrical energy above that frequency?
    A. A band-pass filter
    B. A high-pass filter
    C. An input filter
    D. A low-pass filter

3AG-1-2.1   Why does virtually every modern transmitter have a built-in low-pass filter connected to its output?
    A. To attenuate frequencies below its cutoff point
    B. To attenuate low frequency interference to other amateurs
    C. To attenuate excess harmonic radiation
    D. To attenuate excess fundamental radiation

3AG-1-2.2   You believe that excess harmonic radiation from your transmitter is causing interference to your television receiver. What is one possible solution for this problem?
    A. Install a low-pass filter on the television receiver
    B. Install a low-pass filter at the transmitter output
    C. Install a high-pass filter on the transmitter output
    D. Install a band-pass filter on the television receiver

3AG-2-1.1   What circuit passes electrical energy above a certain frequency and attenuates electrical energy below that frequency?
    A. A band-pass filter
    B. A high-pass filter
    C. An input filter
    D. A low-pass filter

3AG-2-2.1   Where is the proper place to install a high-pass filter?
    A. At the antenna terminals of a television receiver
    B. Between a transmitter and a Transmatch
    C. Between a Transmatch and the transmission line
    D. On a transmitting antenna

3AG-2-2.2   Your Amateur Radio transmissions cause interference to your television receiver even though you have installed a low-pass filter at the transmitter output. What is one possible solution for this problem?
    A. Install a high-pass filter at the transmitter terminals
    B. Install a high-pass filter at the television antenna terminals
    C. Install a low-pass filter at the television antenna terminals also
    D. Install a band-pass filter at the television antenna terminals

3AG-3-1.1   What circuit attenuates electrical energy above a certain frequency and below a lower frequency?
    A. A band-pass filter
    B. A high-pass filter
    C. An input filter
    D. A low-pass filter

3AG-3-1.2   What general range of RF energy does a band-pass filter reject?
  A. All frequencies above a specified frequency
  B. All frequencies below a specified frequency
  C. All frequencies above the upper limit of the band in question
  D. All frequencies above a specified frequency and below a lower specified frequency

3AG-3-2.1   The IF stage of a communications receiver uses a filter with a peak response at the intermediate frequency. What term describes this filter response?
  A. A band-pass filter
  B. A high-pass filter
  C. An input filter
  D. A low-pass filter

3AG-4-1.1   What circuit is likely to be found in all types of receivers?
  A. An audio filter
  B. A beat frequency oscillator
  C. A detector
  D. An RF amplifier

3AG-4-1.2   What type of transmitter does this block diagram represent?

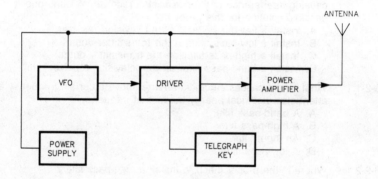

  A. A simple packet-radio transmitter
  B. A simple crystal-controlled transmitter
  C. A single-sideband transmitter
  D. A VFO-controlled transmitter

3AG-4-1.3 What type of transmitter does this block diagram represent?

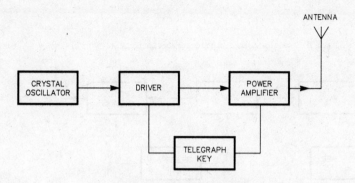

A. A simple packet-radio transmitter
B. A simple crystal-controlled transmitter
C. A single-sideband transmitter
D. A VFO-controlled transmitter

3AG-4-1.4 What is the unlabeled block (?) in this diagram?

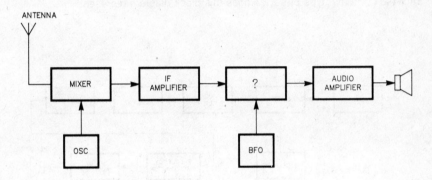

A. An AGC circuit
B. A detector
C. A power supply
D. A VFO circuit

3AG-4-1.5   What type of device does this block diagram represent?

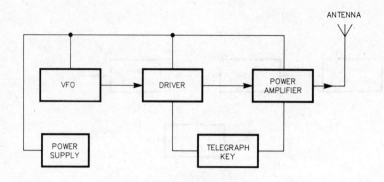

A. A double-conversion receiver
B. A variable-frequency oscillator
C. A simple superheterodyne receiver
D. A simple CW transmitter

3AG-4-2.1   What type of device does this block diagram represent?

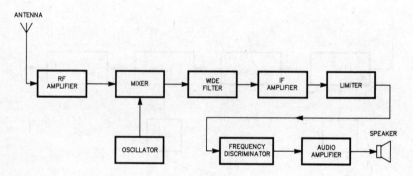

A. A double-conversion receiver
B. A variable-frequency oscillator
C. A simple superheterodyne receiver
D. A simple FM receiver

3AG-4-2.2   What is the unlabeled block (?) in this diagram?

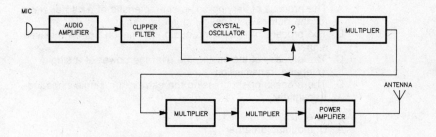

A. A band-pass filter
B. A crystal oscillator
C. A reactance modulator
D. A rectifier modulator

3AH-1.1   What is the meaning of the term *modulation*?
    A. The process of varying some characteristic of a carrier wave for the purpose of conveying information
    B. The process of recovering audio information from a received signal
    C. The process of increasing the average power of a single-sideband transmission
    D. The process of suppressing the carrier in a single-sideband transmitter

3AH-2-1.1   What is emission type NØN?
    A. Unmodulated carrier
    B. Telegraphy by on-off keying
    C. Telegraphy by keyed tone
    D. Telegraphy by frequency-shift keying

3AH-2-1.2   What emission does not have sidebands resulting from modulation?
    A. A3E
    B. NØN
    C. F3E
    D. F2B

3AH-2-2.1   What is the FCC emission designator for a Morse code telegraphy signal produced by switching the transmitter output on and off?
    A. NØN
    B. A3E
    C. A1A
    D. F1B

3AH-2-2.2   What is emission type A1A?
    A. Morse code telegraphy using amplitude modulation
    B. Morse code telegraphy using frequency modulation
    C. Morse code telegraphy using phase modulation
    D. Morse code telegraphy using pulse modulation

3AH-2-3.1   What is emission type F1B?
    A. Amplitude-keyed telegraphy
    B. Frequency-shift-keyed telegraphy
    C. Frequency-modulated telephony
    D. Phase-modulated telephony

3AH-2-3.2   What is the emission symbol for telegraphy by frequency shift keying without the use of a modulating tone?
    A. F1B
    B. F2B
    C. A1A
    D. J3E

3AH-2-4.1   What emission type results when an on/off keyed audio tone is applied to the microphone input of an FM transmitter?
    A. F1B
    B. F2A
    C. A1A
    D. J3E

3AH-2-4.2   What is emission type F2A?
    A. Telephony produced by audio fed into an FM transmitter
    B. Telegraphy produced by an on/off keyed audio tone fed into an AM transmitter
    C. Telegraphy produced by on/off keying of the carrier amplitude
    D. Telegraphy produced by an on/off keyed audio tone fed into an FM transmitter

3AH-2-5.1   What is emission type F2B?
    A. Frequency-modulated telegraphy using audio tones
    B. Frequency-modulated telephony
    C. Frequency-modulated facsimile using audio tones
    D. Phase-modulated television

3AH-2-5.2   What emissions are used in teleprinting?
    A. F1A, F2B and F1B
    B. A2B, F1B and F2B
    C. A1B, A2B and F2B
    D. A2B, F1A and F2B

3AH-2-5.3   What emission type results when an AF shift keyer is connected to the microphone jack of an emission F3E transmitter?
    A. A2B
    B. F1B
    C. F2B
    D. A1F

3AH-2-6.1   What is emission type F2D?
    A. A data transmission produced by modulating an FM transmitter with audio tones
    B. A telemetry transmission produced by modulating an FM transmitter with two sidebands
    C. A data transmission produced by modulating an FM transmitter with pulse modulation
    D. A telemetry transmission produced by modulating an SSB transmitter with phase modulation

3AH-2-6.2   What FCC emission designator describes a packet-radio transmission through an FM transmitter?
    A. F1D
    B. F2D
    C. F2B
    D. F1B

3AH-2-7.1   What is emission type F3E?
    A. AM telephony
    B. AM telegraphy
    C. FM telegraphy
    D. FM telephony

3AH-2-7.2   What is the emission symbol for telephony by frequency modulation?
    A. F2B
    B. F3E
    C. A3E
    D. F1B

3AH-2-8.1 What is the FCC emission designator for telephony by phase modulation?
- A. J3E
- B. F1B
- C. G3E
- D. F3E

3AH-2-8.2 What is emission type G3E?
- A. Phase-modulated telegraphy
- B. Frequency-modulated telegraphy
- C. Frequency-modulated telephony
- D. Phase-modulated telephony

3AH-3.1 What is the term used to describe a constant-amplitude radio-frequency signal?
- A. An RF carrier
- B. An AF carrier
- C. A sideband carrier
- D. A subcarrier

3AH-3.2 What is another name for an unmodulated radio-frequency signal?
- A. An AF carrier
- B. An RF carrier
- C. A sideband carrier
- D. A subcarrier

3AH-4.1 What characteristic makes emission F3E especially well-suited for local VHF/UHF radio communications?
- A. Good audio fidelity and intelligibility under weak-signal conditions
- B. Better rejection of multipath distortion than the AM modes
- C. Good audio fidelity and high signal-to-noise ratio above a certain signal amplitude threshold
- D. Better carrier frequency stability than the AM modes

3AH-5.1 What emission is produced by a transmitter using a reactance modulator?
- A. A1A
- B. NØN
- C. J3E
- D. G3E

3AH-5.2 What other emission does phase modulation most resemble?
- A. Amplitude modulation
- B. Pulse modulation
- C. Frequency modulation
- D. Single-sideband modulation

3AH-6.1 Many communications receivers have several IF filters that can be selected by the operator. Why do these filters have different bandwidths?
- A. Because some ham bands are wider than others
- B. Because different bandwidths help increase the receiver sensitivity
- C. Because different bandwidths improve S-meter readings
- D. Because some emission types occupy a wider frequency range than others

3AH-6.2 List the following signals in order of increasing bandwidth (narrowest signal first): CW, FM voice, RTTY, SSB voice.
A. RTTY, CW, SSB voice, FM voice
B. CW, FM voice, RTTY, SSB voice
C. CW, RTTY, SSB voice, FM voice
D. CW, SSB voice, RTTY, FM voice

3AH-7-1.1 To what is the deviation of an emission F3E transmission proportional?
A. Only the frequency of the audio modulating signal
B. The frequency and the amplitude of the audio modulating signal
C. The duty cycle of the audio modulating signal
D. Only the amplitude of the audio modulating signal

3AH-7-2.1 What is the result of overdeviation in an emission F3E transmitter?
A. Increased transmitter power consumption
B. Out-of-channel emissions (splatter)
C. Increased transmitter range
D. Inadequate carrier suppression

3AH-7-2.2 What is *splatter*?
A. Interference to adjacent signals caused by excessive transmitter keying speeds
B. Interference to adjacent signals caused by improper transmitter neutralization
C. Interference to adjacent signals caused by overmodulation of a transmitter
D. Interference to adjacent signals caused by parasitic oscillations at the antenna

SUBELEMENT 3AI—Antennas and Feed Lines (3 Exam Questions)

3AI-1-1.1   What antenna type best strengthens signals from a particular direction while attenuating those from other directions?
   A. A beam antenna
   B. An isotropic antenna
   C. A monopole antenna
   D. A vertical antenna

3AI-1-1.2   What is a *directional antenna*?
   A. An antenna whose parasitic elements are all constructed to be directors
   B. An antenna that radiates in direct line-of-sight propagation, but not sky-wave or skip propagation
   C. An antenna permanently mounted so as to radiate in only one direction
   D. An antenna that radiates more strongly in some directions than others

3AI-1-1.3   What is a *Yagi* antenna?
   A. Half-wavelength elements stacked vertically and excited in phase
   B. Quarter-wavelength elements arranged horizontally and excited out of phase
   C. Half-wavelength linear driven element(s) with parasitically excited parallel linear elements
   D. Quarter-wavelength, triangular loop elements

3AI-1-1.4   What is the general configuration of the radiating elements of a horizontally polarized Yagi?
   A. Two or more straight, parallel elements arranged in the same horizontal plane
   B. Vertically stacked square or circular loops arranged in parallel horizontal planes
   C. Two or more wire loops arranged in parallel vertical planes
   D. A vertical radiator arranged in the center of an effective RF ground plane

3AI-1-1.5   What type of parasitic beam antenna uses two or more straight metal-tubing elements arranged physically parallel to each other?
   A. A delta loop antenna
   B. A quad antenna
   C. A Yagi antenna
   D. A Zepp antenna

3AI-1-1.6   How many directly driven elements does a Yagi antenna have?
   A. None; they are all parasitic
   B. One
   C. Two
   D. All elements are directly driven

3AI-1-1.7   What is a *parasitic beam antenna?*
- A. An antenna where the director and reflector elements receive their RF excitation by induction or radiation from the driven element
- B. An antenna where wave traps are used to assure magnetic coupling among the elements
- C. An antenna where all elements are driven by direct connection to the feed line
- D. An antenna where the driven element receives its RF excitation by induction or radiation from the directors

3AI-1-2.1   What is a *cubical quad antenna?*
- A. Four parallel metal tubes, each approximately 1/2 electrical wavelength long
- B. Two or more parallel four-sided wire loops, each approximately one electrical wavelength long
- C. A vertical conductor 1/4 electrical wavelength high, fed at the bottom
- D. A center-fed wire 1/2 electrical wavelength long

3AI-1-2.2   What kind of antenna array is composed of a square full-wave closed loop driven element with parallel parasitic element(s)?
- A. Delta loop
- B. Cubical quad
- C. Dual rhombic
- D. Stacked Yagi

3AI-1-2.3   Approximately how long is one side of the driven element of a cubical quad antenna?
- A. 2 electrical wavelengths
- B. 1 electrical wavelength
- C. 1/2 electrical wavelength
- D. 1/4 electrical wavelength

3AI-1-2.4   Approximately how long is the wire in the driven element of a cubical quad antenna?
- A. 1/4 electrical wavelength
- B. 1/2 electrical wavelength
- C. 1 electrical wavelength
- D. 2 electrical wavelengths

3AI-1-3.1   What is a *delta loop antenna?*
- A. A variation of the cubical quad antenna, with triangular elements
- B. A large copper ring, used in direction finding
- C. An antenna system composed of three vertical antennas, arranged in a triangular shape
- D. An antenna made from several coils of wire on an insulating form

3AI-2-1.1   To what does the term *horizontal* as applied to wave polarization refer?
- A. The magnetic lines of force in the radio wave are parallel to the earth's surface
- B. The electric lines of force in the radio wave are parallel to the earth's surface
- C. The electric lines of force in the radio wave are perpendicular to the earth's surface
- D. The radio wave will leave the antenna and radiate horizontally to the destination

3AI-2-1.2   What electromagnetic wave polarization does a cubical quad
             antenna have when the feed point is in the center of a horizontal
             side?
             A. Circular
             B. Helical
             C. Horizontal
             D. Vertical

3AI-2-1.3   What electromagnetic wave polarization does a cubical quad
             antenna have when all sides are at 45 degrees to the earth's
             surface and the feed point is at the bottom corner?
             A. Circular
             B. Helical
             C. Horizontal
             D. Vertical

3AI-2-2.1   What is the polarization of electromagnetic waves radiated from a
             half-wavelength antenna perpendicular to the earth's surface?
             A. Circularly polarized waves
             B. Horizontally polarized waves
             C. Parabolically polarized waves
             D. Vertically polarized waves

3AI-2-2.2   What is the electromagnetic wave polarization of most man-made
             electrical noise radiation in the HF-VHF spectrum?
             A. Horizontal
             B. Left-hand circular
             C. Right-hand circular
             D. Vertical

3AI-2-2.3   To what does the term *vertical* as applied to wave polarization refer?
             A. The electric lines of force in the radio wave are parallel to the
                earth's surface
             B. The magnetic lines of force in the radio wave are
                perpendicular to the earth's surface
             C. The electric lines of force in the radio wave are perpendicular
                to the earth's surface
             D. The radio wave will leave the antenna and radiate vertically
                into the ionosphere

3AI-2-2.4   What electromagnetic wave polarization does a cubical quad
             antenna have when the feed point is in the center of a vertical side?
             A. Circular
             B. Helical
             C. Horizontal
             D. Vertical

3AI-2-2.5   What electromagnetic wave polarization does a cubical quad
             antenna have when all sides are at 45 degrees to the earth's
             surface and the feed point is at a side corner?
             A. Circular
             B. Helical
             C. Horizontal
             D. Vertical

3AI-3-1.1   What is meant by the term *standing wave ratio*?
             A. The ratio of maximum to minimum inductances on a feed line
             B. The ratio of maximum to minimum resistances on a feed line
             C. The ratio of maximum to minimum impedances on a feed line
             D. The ratio of maximum to minimum voltages on a feed line

3AI-3-1.2   What is *standing wave ratio* a measure of?
- A. The ratio of maximum to minimum voltage on a feed line
- B. The ratio of maximum to minimum reactance on a feed line
- C. The ratio of maximum to minimum resistance on a feed line
- D. The ratio of maximum to minimum sidebands on a feed line

3AI-3-2.1   What is meant by the term *forward power*?
- A. The power traveling from the transmitter to the antenna
- B. The power radiated from the front of a directional antenna
- C. The power produced during the positive half of the RF cycle
- D. The power used to drive a linear amplifier

3AI-3-2.2   What is meant by the term *reflected power*?
- A. The power radiated from the back of a directional antenna
- B. The power returned to the transmitter from the antenna
- C. The power produced during the negative half of the RF cycle
- D. Power reflected to the transmitter site by buildings and trees

3AI-3-3.1   What happens to the power loss in an unbalanced feed line as the standing wave ratio increases?
- A. It is unpredictable
- B. It becomes nonexistent
- C. It decreases
- D. It increases

3AI-3-3.2   What type of feed line is best suited to operating at a high standing wave ratio?
- A. Coaxial cable
- B. Flat ribbon "twin lead"
- C. Parallel open-wire line
- D. Twisted pair

3AI-3-3.3   What happens to RF energy not delivered to the antenna by a lossy coaxial cable?
- A. It is radiated by the feed line
- B. It is returned to the transmitter's chassis ground
- C. Some of it is dissipated as heat in the conductors and dielectric
- D. It is canceled because of the voltage ratio of forward power to reflected power in the feed line

3AI-4-1.1   What is a *balanced line*?
- A. Feed line with one conductor connected to ground
- B. Feed line with both conductors connected to ground to balance out harmonics
- C. Feed line with the outer conductor connected to ground at even intervals
- D. Feed line with neither conductor connected to ground

3AI-4-1.2   What is an *unbalanced line*?
- A. Feed line with neither conductor connected to ground
- B. Feed line with both conductors connected to ground to suppress harmonics
- C. Feed line with one conductor connected to ground
- D. Feed line with the outer conductor connected to ground at uneven intervals

3AI-4-2.1   What is a *balanced antenna*?
  A. A symmetrical antenna with one side of the feed point connected to ground
  B. An antenna (or a driven element in an array) that is symmetrical about the feed point
  C. A symmetrical antenna with both sides of the feed point connected to ground, to balance out harmonics
  D. An antenna designed to be mounted in the center

3AI-4-2.2   What is an *unbalanced antenna*?
  A. An antenna (or a driven element in an array) that is not symmetrical about the feed point
  B. A symmetrical antenna, having neither half connected to ground
  C. An antenna (or a driven element in an array) that is symmetrical about the feed point
  D. A symmetrical antenna with both halves coupled to ground at uneven intervals

3AI-4-3.1   What device can be installed on a balanced antenna so that it can be fed through a coaxial cable?
  A. A balun
  B. A loading coil
  C. A triaxial transformer
  D. A wavetrap

3AI-4-3.2   What is a *balun*?
  A. A device that can be used to convert an antenna designed to be fed at the center so that it may be fed at one end
  B. A device that may be installed on a balanced antenna so that it may be fed with unbalanced feed line
  C. A device that can be installed on an antenna to produce horizontally polarized or vertically polarized waves
  D. A device used to allow an antenna to operate on more than one band

3AI-5-1.1   List the following types of feed line in order of increasing attenuation per 100 feet of line (list the line with the lowest attenuation first): RG-8, RG-58, RG-174 and open-wire line.
  A. RG-174, RG-58, RG-8, open-wire line
  B. RG-8, open-wire line, RG-58, RG-174
  C. open-wire line, RG-8, RG-58, RG-174
  D. open-wire line, RG-174, RG-58, RG-8

3AI-5-1.2   You have installed a tower 150 feet from your radio shack, and have a 6-meter Yagi antenna on top. Which of the following feed lines should you choose to feed this antenna: RG-8, RG-58, RG-59 or RG-174?
  A. RG-8
  B. RG-58
  C. RG-59
  D. RG-174

3AI-5-2.1   You have a 200-foot coil of RG-58 coaxial cable attached to your
             antenna, but the antenna is only 50 feet from your radio. To
             minimize feed-line loss, what should you do with the excess cable?
             A. Cut off the excess cable to an even number of wavelengths
                long
             B. Cut off the excess cable to an odd number of wavelengths
                long
             C. Cut off the excess cable
             D. Roll the excess cable into a coil a tenth of a wavelength in
                diameter

3AI-5-2.2   How does feed-line length affect signal loss?
             A. The length has no effect on signal loss
             B. As length increases, signal loss increases
             C. As length decreases, signal loss increases
             D. The length is inversely proportional to signal loss

3AI-5-3.1   What is the general relationship between frequencies passing
             through a feed line and the losses in the feed line?
             A. Loss is independent of frequency
             B. Loss increases with increasing frequency
             C. Loss decreases with increasing frequency
             D. There is no predictable relationship

3AI-5-3.2   As the operating frequency decreases, what happens to conductor
             losses in a feed line?
             A. The losses decrease
             B. The losses increase
             C. The losses remain the same
             D. The losses become infinite

3AI-5-3.3   As the operating frequency increases, what happens to conductor
             losses in a feed line?
             A. The losses decrease
             B. The losses increase
             C. The losses remain the same
             D. The losses decrease to zero

3AI-6-1.1   You are using open-wire feed line in your Amateur Radio station.
             Why should you ensure that no one can come in contact with the
             feed line while you are transmitting?
             A. Because contact with the feed line while tranmitting will cause
                a short circuit, probably damaging your transmitter
             B. Because the wire is so small they may break it
             C. Because contact with the feed line while transmitting will
                cause parasitic radiation
             D. Because high RF voltages can be present on open-wire feed
                line

3AI-6-2.1   How can you minimize exposure to radio frequency energy from
             your transmitting antennas?
             A. Use vertical polarization
             B. Use horizontal polarization
             C. Mount the antennas where no one can come near them
             D. Mount the antenna close to the ground

ELEMENT 3A ANSWER KEY

SUBELEMENT 2A

Numbers in this section refer to FCC Rules, Part 97. These references are from the Rules in effect when the question pool was revised. Rewritten FCC Rules went into effect September 1, 1989. The VECs are expected to release a supplement to update the questions around March 1, 1990.

3AA-1.1	A	{97.3(p)}
3AA-1.2	B	{97.3(p)}
3AA-2.1	B	{97.7(b),(f)}
3AA-2.2	C	{97.7(b),(f)}
3AA-2.3	B	{97.7(b),(f)}
3AA-2.4	A	{97.7(b),(f)}
3AA-2.5	B	{97.7(b),(f)}
3AA-3.1	A	{97.13; 97.47}
3AA-3.2	A	{97.13; 97.47}
3AA-3.3	A	{97.13; 97.47}
3AA-4.1	B	{97.61}
3AA-4.2	A	{97.61}
3AA-4.3	A	{97.61}
3AA-5.1	D	{97.63}
3AA-5.2	C	{97.63}
3AA-6-1.1	C	{97.3(t)(1)}
3AA-6-1.2	D	{97.3(t)(1)}
3AA-6-2.1	C	{97.67(a)}
3AA-6-3.1	D	{97.67(b),(d)}
3AA-6-4.1	B	{97.67(e)}
3AA-7-1.1	C	{97.69(a)(1)}
3AA-7-1.2	B	{97.69(a)(1)}
3AA-7-1.3	D	{97.69(a)(1)}
3AA-7-2.1	C	{97.69(a)(2)}
3AA-7-2.2	C	{97.69(a)(2)}
3AA-7-3.1	A	{97.69(c)(2)}
3AA-7-3.2	D	{97.69(c)(2)}
3AA-7-3.3	D	{97.69(c)(2)}
3AA-8-1.1	B	{97.35; 97.84(f)}
3AA-8-2.1	B	{97.84(g)(2)}
3AA-8-3.1	C	{97.84(g)(2)}
3AA-9-1.1	A	{97.3(l)}
3AA-9-2.1	A	{97.87(f)}
3AA-10.1	A	{97.3(l); 97.99}
3AA-10.2	C	{97.3(l); 97.99}
3AA-10.3	D	{97.3(l); 97.99}

3AA-10.4	B	{97.3(l); 97.99}
3AA-11-1.1	A	{97.3(w)}
3AA-11-1.2	B	{97.3(w)}
3AA-11-1.3	A	{97.3(w)}
3AA-11-2.1	D	{97.107}
3AA-11-2.2	A	{97.107}
3AA-11-2.3	C	{97.107}
3AA-11-2.4	A	{97.107}
3AA-12.1	A	{97.113(a), (b), (c)}
3AA-12.2	C	{97.113(a), (b), (c)}
3AA-12.3	B	{97.113(a), (b), (c)}
3AA-12.4	D	{97.113(a), (b), (c)}
3AA-12.5	C	{97.113(a), (b), (c)}
3AA-13.1	B	{97.113(d)}
3AA-13.2	D	{97.113(d)}
3AA-13.3	D	{97.113(d)}
3AA-13.4	C	{97.113(d)}
3AA-14.1	D	{97.114(a), (b)}
3AA-14.2	C	{97.114(a), (b)}
3AA-14.3	D	{97.114(a), (b)}
3AA-15.1	A	{97.114(c)}
3AA-15.2	C	{97.114(c)}
3AA-15.3	D	{97.114(c)}
3AA-15.4	B	{97.114(c)}
3AA-16.1	B	{97.119}
3AA-16.2	D	{97.119}
3AA-16.3	C	{97.119}
3AA-17.1	A	{97.409}

SUBELEMENT 3AB

Numbers in this section refer to page numbers in *The ARRL Technician Class License Manual.*

3AB-1.1	A	p 2-2
3AB-1.2	C	p 2-2
3AB-1.3	D	p 2-2
3AB-2-1.1	B	p 2-4
3AB-2-1.2	C	p 2-4
3AB-2-1.3	A	p 2-4
3AB-2-1.4	D	p 2-4
3AB-2-1.5	B	p 2-3
3AB-2-1.6	B	p 2-4
3AB-2-1.7	D	p 2-6

3AB-2-2.1	C	p 2-4		3AC-7.3	D	p 3-10
3AB-2-2.2	C	p 2-6		3AC-7.4	A	p 3-10
3AB-2-3.1	D	p 2-5		3AC-7.5	B	p 3-10
3AB-2-3.2	B	p 2-6		3AC-7.6	D	p 3-11
3AB-2-3.3	A	p 2-5				

3AB-2-3.4	C	p 2-6
3AB-2-4.1	D	p 2-4
3AB-3.1	A	p 2-6
3AB-3.2	B	p 2-6
3AB-3.3	C	p 2-6
3AB-4.1	A	p 2-7
3AB-4.2	D	p 2-7
3AB-5-1.1	C	p 2-7
3AB-5-1.2	B	p 2-7
3AB-5-2.1	D	p 2-7
3AB-6-1.1	A	p 2-7
3AB-6-1.2	B	p 2-8
3AB-6-2.1	D	p 2-8
3AB-6-3.1	B	p 2-8
3AB-6-3.2	C	p 2-8

SUBELEMENT 3AC

3AC-1-1.1	A	p 3-2
3AC-1-1.2	D	p 3-1
3AC-1-1.3	C	p 3-2
3AC-1-2.1	A	p 3-2
3AC-1-2.2	B	p 3-2
3AC-1-3.1	B	p 3-4
3AC-1-4.1	D	p 3-4
3AC-1-4.2	B	p 3-4
3AC-1-4.3	C	p 3-4
3AC-2.1	D	p 3-2
3AC-2.2	B	p 3-2
3AC-2.3	A	p 3-3
3AC-2.4	B	p 3-2
3AC-3.1	D	p 3-4
3AC-3.2	C	p 3-4
3AC-3.3	A	p 3-4
3AC-3.4	B	p 3-4
3AC-4.1	D	p 3-5
3AC-4.2	C	p 3-5
3AC-4.3	A	p 3-5
3AC-5.1	C	p 3-7
3AC-5.2	C	p 3-9
3AC-6.1	A	p 3-9
3AC-6.2	B	p 3-9
3AC-7.1	C	p 3-10
3AC-7.2	A	p 3-10

SUBELEMENT 3AD

3AD-1-1.1	C	p 4-3
3AD-1-1.2	D	p 4-3
3AD-1-1.3	B	p 4-3
3AD-1-1.4	B	p 4-4
3AD-1-2.1	A	p 4-5
3AD-1-2.2	A	p 4-5
3AD-1-2.3	A	p 4-5
3AD-1-3.1	C	p 4-5
3AD-2-1.1	B	p 4-7
3AD-2-2.1	C	p 4-8
3AD-3-1.1	A	p 4-8
3AD-3-2.1	D	p 4-8
3AD-4.1	D	p 4-8
3AD-5-1.1	A	p 4-10
3AD-5-1.2	C	p 4-10
3AD-5-1.3	B	p 4-10
3AD-5-1.4	A	p 4-10
3AD-5-2.1	B	p 4-10
3AD-5-2.2	C	p 4-10
3AD-6.1	A	p 4-11
3AD-6.2	D	p 4-11
3AD-6.3	A	p 4-11
3AD-7.1	C	p 4-11
3AD-7.2	B	p 4-11
3AD-8-1.1	D	p 4-12
3AD-8-1.2	D	p 4-12
3AD-8-2.1	A	p 4-12
3AD-8-2.2	C	p 4-12
3AD-9.1	D	p 4-13
3AD-9.2	B	p 4-13
3AD-9.3	C	p 4-13
3AD-9.4	B	p 4-13
3AD-9.5	A	p 4-13
3AD-9.6	A	p 4-13
3AD-10.1	C	p 4-13
3AD-10.2	A	p 4-13
3AD-11-1.1	B	p 4-14
3AD-11-1.2	A	p 4-14
3AD-11-2.1	D	p 4-14
3AD-11-2.2	B	p 4-14
3AD-11-2.3	B	p 4-14
3AD-11-2.4	D	p 4-14

SUBELEMENT 3AH

3AH-1.1	A	p 8-1
3AH-2-1.1	A	p 8-2
3AH-2-1.2	B	p 8-3
3AH-2-2.1	C	p 8-2
3AH-2-2.2	A	p 8-2
3AH-2-3.1	B	p 8-2
3AH-2-3.2	A	p 8-2
3AH-2-4.1	B	p 8-2
3AH-2-4.2	D	p 8-2
3AH-2-5.1	A	p 8-2
3AH-2-5.2	B	p 8-3
3AH-2-5.3	C	p 8-3
3AH-2-6.1	A	p 8-2
3AH-2-6.2	B	p 8-3
3AH-2-7.1	D	p 8-2
3AH-2-7.2	B	p 8-2
3AH-2-8.1	C	p 8-2
3AH-2-8.2	D	p 8-2
3AH-3.1	A	p 8-5
3AH-3.2	B	p 8-5
3AH-4.1	C	p 8-3
3AH-5.1	D	p 8-4
3AH-5.2	C	p 8-4
3AH-6.1	D	p 8-6
3AH-6.2	C	p 8-6
3AH-7-1.1	D	p 8-7
3AH-7-2.1	B	p 8-8
3AH-7-2.2	C	p 8-8

SUBELEMENT 3AI

3AI-1-1.1	A	p 9-2
3AI-1-1.2	D	p 9-2
3AI-1-1.3	C	p 9-3
3AI-1-1.4	A	p 9-5
3AI-1-1.5	C	p 9-3
3AI-1-1.6	B	p 9-3
3AI-1-1.7	A	p 9-2
3AI-1-2.1	B	p 9-5
3AI-1-2.2	B	p 9-5
3AI-1-2.3	D	p 9-6
3AI-1-2.4	C	p 9-6
3AI-1-3.1	A	p 9-7
3AI-2-1.1	B	p 9-9
3AI-2-1.2	C	p 9-6
3AI-2-1.3	C	p 9-6
3AI-2-2.1	D	p 9-9
3AI-2-2.2	D	p 9-9
3AI-2-2.3	C	p 9-9
3AI-2-2.4	D	p 9-6
3AI-2-2.5	D	p 9-6
3AI-3-1.1	D	p 9-9
3AI-3-1.2	A	p 9-10
3AI-3-2.1	A	p 9-10
3AI-3-2.2	B	p 9-10
3AI-3-3.1	D	p 9-10
3AI-3-3.2	C	p 9-11
3AI-3-3.3	C	p 9-14
3AI-4-1.1	D	p 9-11
3AI-4-1.2	C	p 9-11
3AI-4-2.1	B	p 9-11
3AI-4-2.2	A	p 9-11
3AI-4-3.1	A	p 9-11
3AI-4-3.2	B	p 9-11
3AI-5-1.1	C	p 9-14
3AI-5-1.2	A	p 9-14
3AI-5-2.1	C	p 9-15
3AI-5-2.2	B	p 9-15
3AI-5-3.1	B	p 9-13
3AI-5-3.2	A	p 9-13
3AI-5-3.3	B	p 9-13
3AI-6-1.1	D	p 9-15
3AI-6-2.1	C	p 9-15

Appendix A

Useful Tables

US Customary—Metric Conversion Factors

International System of Units (SI) — Metric Units

Prefix	Symbol	Multiplication Factor	
exa	E	10^{18} =	1,000,000,000,000,000,000
peta	P	10^{15} =	1,000,000,000,000,000
tera	T	10^{12} =	1,000,000,000,000
giga	G	10^{9} =	1,000,000,000
mega	M	10^{6} =	1,000,000
kilo	k	10^{3} =	1,000
hecto	h	10^{2} =	100
deca	da	10^{1} =	10
(unit)		10^{0} =	1
deci	d	10^{-1} =	0.1
centi	c	10^{-2} =	0.01
milli	m	10^{-3} =	0.001
micro	μ	10^{-6} =	0.000001
nano	n	10^{-9} =	0.000000001
pico	p	10^{-12} =	0.000000000001
femto	f	10^{-15} =	0.000000000000001
atto	a	10^{-18} =	0.000000000000000001

Linear
1 meter (m) = 100 centimeters (cm) = 1000 millimeters (mm)

Area
$1\ m^2 = 1 \times 10^4\ cm^2 = 1 \times 10^6\ mm^2$

Volume
$1\ m^3 = 1 \times 10^6\ cm^3 = 1 \times 10^9\ mm^3$
1 liter (l) = 1000 cm^3 = 1 × 10^6 mm^3

Mass
1 kilogram (kg) = 1000 grams (g)
(Approximately the mass of 1 liter of water)
1 metric ton (or tonne) = 1000 kg

US Customary Units

Linear Units
12 inches (in) = 1 foot (ft)
36 inches = 3 feet = 1 yard (yd)
1 rod = 5½ yards = 16½ feet
1 statute mile = 1760 yards = 5280 feet
1 nautical mile = 6076.11549 feet

Area
$1\ ft^2 = 144\ in^2$
$1\ yd^2 = 9\ ft^2 = 1296\ in^2$
$1\ rod^2 = 30\frac{1}{4}\ yd^2$
$1\ acre = 4840\ yd^2 = 43{,}560\ ft^2$
$1\ acre = 160\ rod^2$
$1\ mile^2 = 640\ acres$

Volume
$1\ ft^3 = 1728\ in^3$
$1\ yd^3 = 27\ ft^3$

Liquid Volume Measure
1 fluid ounce (fl oz) = 8 fluidrams = 1.804 in^3
1 pint (pt) = 16 fl oz
1 quart (qt) = 2 pt = 32 fl oz = 57¾ in^3
1 gallon (gal) = 4 qt = 231 in^3
1 barrel = 31½ gal

Dry Volume Measure
1 quart (qt) = 2 pints (pt) = 67.2 in^3
1 peck = 8 qt
1 bushel = 4 pecks = 2150.42 in^3

Avoirdupois Weight
1 dram (dr) = 27.343 grains (gr) or (gr a)
1 ounce (oz) = 437.5 gr
1 pound (lb) = 16 oz = 7000 gr
1 short ton = 2000 lb, 1 long ton = 2240 lb

Troy Weight
1 grain troy (gr t) = 1 grain avoirdupois
1 pennyweight (dwt) or (pwt) = 24 gr t
1 ounce troy (oz t) = 480 grains
1 lb t = 12 oz t = 5760 grains

Apothecaries' Weight
1 grain apothecaries' (gr ap) = 1 gr t = 1 gr a

1 dram ap (dr ap) = 60 gr
1 oz ap = 1 oz t = 8 dr ap = 480 gr
1 lb ap = 1 lb t = 12 oz ap = 5760 gr

Temperature
°F = 9/5°C + 32
°C = 5/9 (°F − 32)
K = °C + 273
°C = K − 273

Multiply ⟶

Metric Unit = Conversion Factor × U.S. Customary Unit

◀— Divide

Metric Unit ÷ Conversion Factor = U.S. Customary Unit

	Conversion				Conversion	
Metric Unit	= Factor ×	U.S. Unit		Metric Unit	= Factor ×	U.S. Unit
(Length)				**(Volume)**		
mm	25.4	inch		mm^3	16387.064	in^3
cm	2.54	inch		cm^3	16.387	in^3
cm	30.48	foot		m^3	0.028316	ft^3
m	0.3048	foot		m^3	0.764555	yd^3
m	0.9144	yard		ml	16.387	in^3
km	1.609	mile		ml	29.57	fl oz
km	1.852	nautical mile		ml	473	pint
(Area)				ml	946.333	quart
mm^2	645.16	$inch^2$		l	28.32	ft^3
cm^2	6.4516	in^2		l	0.9463	quart
cm^2	929.03	ft^2		l	3.785	gallon
m^2	0.0929	ft^2		l	1.101	dry quart
cm^2	8361.3	yd^2		l	8.809	peck
m^2	0.83613	yd^2		l	35.238	bushel
m^2	4047	acre				
km^2	2.59	mi^2				
(Mass)	**(Avoirdupois Weight)**			**(Mass)**	**(Troy Weight)**	
grams	0.0648	grains		g	31.103	oz t
g	28.349	oz		g	373.248	lb t
g	453.59	lb		**(Mass)**	**(Apothecaries' Weight)**	
kg	0.45359	lb		g	3.387	dr ap
tonne	0.907	short ton		g	31.103	oz ap
tonne	1.016	long ton		g	373.248	lb ap

Standard Resistance Values

Numbers in **bold** type are ±10% values. Others are 5% values.

Ohms											Megohms				
1.0	3.6	**12**	43	**150**	510	**1800**	6200	**22000**	75000		0.24	0.62	1.6	4.3	11.0
1.1	**3.9**	13	**47**	160	**560**	2000	**6800**	24000	**82000**		**0.27**	**0.68**	1.8	**4.7**	12.0
1.2	4.3	**15**	51	**180**	620	**2200**	7500	**27000**	91000		0.30	0.75	2.0	5.1	13.0
1.3	**4.7**	16	**56**	200	**680**	2400	**8200**	30000	**100000**		**0.33**	**0.82**	2.2	5.6	**15.0**
1.5	5.1	**18**	62	**220**	750	**2700**	9100	**33000**	110000		0.36	0.91	2.4	6.2	16.0
1.6	**5.6**	20	**68**	240	**820**	3000	**10000**	36000	**120000**		**0.39**	**1.0**	2.7	**6.8**	18.0
1.8	6.2	**22**	75	**270**	910	**3300**	11000	**39000**	130000		0.43	1.1	3.0	7.5	20.0
2.0	**6.8**	24	**82**	300	**1000**	3600	**12000**	43000	**150000**		**0.47**	**1.2**	3.3	**8.2**	22.0
2.2	7.5	**27**	91	**330**	1100	**3900**	13000	**47000**	160000		0.51	1.3	3.6	9.1	
2.4	**8.2**	30	**100**	360	**1200**	4300	**15000**	51000	**180000**		**0.56**	**1.5**	3.9	**10.0**	
2.7	9.1	**33**	110	**390**	1300	**4700**	16000	**56000**	200000						
3.0	**10.0**	36	**120**	430	**1500**	5100	**18000**	62000	**220000**						
3.3	11.0	**39**	130	**470**	1600	**5600**	20000	**68000**							

Resistor Color Code

Color	Sig. Figure	Decimal Multiplier	Tolerance (%)	Color	Sig. Figure	Decimal Multiplier	Tolerance (%)
Black	0	1		Violet	7	10,000,000	
Brown	1	10		Gray	8	100,000,000	
Red	2	100		White	9	1,000,000,000	
Orange	3	1,000		Gold	—	0.1	5
Yellow	4	10,000		Silver	—	0.01	10
Green	5	100,000		No color	—		20
Blue	6	1,000,000					

Schematic Symbols Used in Circuit Diagrams

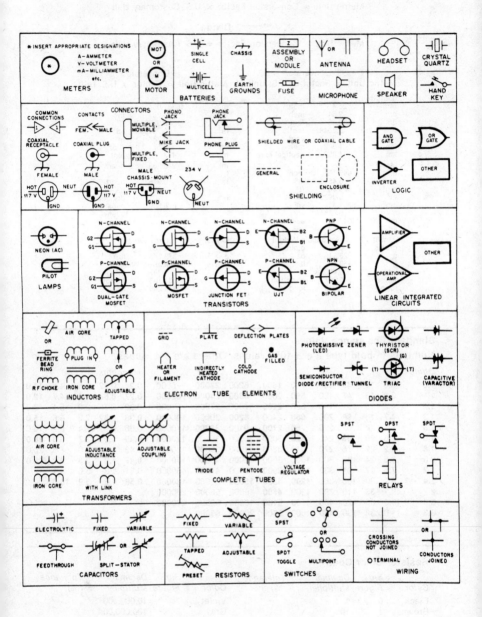

Standard Capacitance Values

pF	pF
0.3	470
5	500
6	510
6.8	560
7.5	600
8	680
10	750
12	800
15	820
18	910
20	1000
22	1000
24	1200
25	1200
27	1300
30	1500
33	1500
39	1600
47	1800
50	2000
51	2200
56	2500
68	2700
75	3000
82	3300
91	3900
100	4000
120	4300
130	4700
150	4700
180	5000
200	5000
220	5600
240	6800
250	7500
270	8200
300	10000
330	10000
350	20000
360	30000
390	40000
400	50000
470	

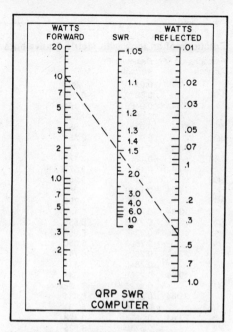

Nomograph of SWR versus forward and reflected power for levels up to 20 watts. Dashed line shows an SWR of 1.5:1 for 10 W forward and 0.4 W reflected.

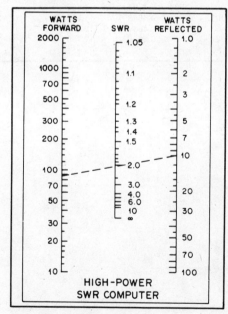

Nomograph of SWR versus forward and reflected power for levels up to 2000 watts. Dashed line shows an SWR of 2:1 for 90 W forward and 10 W reflected.

Fractions of an Inch with Metric Equivalents

Fractions Of An Inch		Decimals Of An Inch	Millimeters	Fractions Of An Inch		Decimals Of An Inch	Millimeters
	1/64	0.0156	0.397		33/64	0.5156	13.097
1/32		0.0313	0.794	17/32		0.5313	13.494
	3/64	0.0469	1.191		35/64	0.5469	13.891
		0.0625	1.588	9/16		0.5625	14.288
	5/64	0.0781	1.984		37/64	0.5781	14.684
3/32		0.0938	2.381	19/32		0.5938	15.081
	7/64	0.1094	2.778		39/64	0.6094	15.478
1/8		0.1250	3.175	5/8		0.6250	15.875
	9/64	0.1406	3.572		41/64	0.6406	16.272
5/32		0.1563	3.969	21/32		0.6563	16.669
	11/64	0.1719	4.366		43/64	0.6719	17.066
3/16		0.1875	4.763	11/16		0.6875	17.463
	13/64	0.2031	5.159		45/64	0.7031	17.859
7/32		0.2188	5.556	23/32		0.7188	18.256
	15/64	0.2344	5.953		47/64	0.7344	18.653
1/4		0.2500	6.350	3/4		0.7500	19.050
	17/64	0.2656	6.747		49/64	0.7656	19.447
9/32		0.2813	7.144	25/32		0.7813	19.844
	19/64	0.2969	7.541		51/64	0.7969	20.241
5/16		0.3125	7.938	13/16		0.8125	20.638
	21/64	0.3281	8.334		53/64	0.8281	21.034
11/32		0.3438	8.731	27/32		0.8438	21.431
	23/64	0.3594	9.128		55/64	0.8594	21.828
3/8		0.3750	9.525	7/8		0.8750	22.225
	25/64	0.3906	9.922		57/64	0.8906	22.622
13/32		0.4063	10.319	29/32		0.9063	23.019
	27/64	0.4219	10.716		59/64	0.9219	23.416
7/16		0.4375	11.113	15/16		0.9375	23.813
	29/64	0.4531	11.509		61/64	0.9531	24.209
15/32		0.4688	11.906	31/32		0.9688	24.606
	31/64	0.4844	12.303		63/64	0.9844	25.003
1/2		0.50000	12.700	1		1.0000	25.400

Appendix B

Equations Used in this Book

True forward power = Forward power reading − Reflected power reading

(Equation 4-1)

$$R_{TOTAL} = R_1 + R_2 + R_3 + \ldots + R_n \qquad \text{(Equation 5-1)}$$

$$R_{TOTAL} = \cfrac{1}{\cfrac{1}{R_1} + \cfrac{1}{R_2} + \cfrac{1}{R_3} + \ldots + \cfrac{1}{R_n}} \qquad \text{(Equation 5-2)}$$

$$R_{TOTAL} = \frac{R_1 \times R_2}{R_1 + R_2} \qquad \text{(Equation 5-3)}$$

$$E = I \times R \qquad \text{(Equation 5-4)}$$

$$I = \frac{E}{R} \qquad \text{(Equation 5-5)}$$

$$R = \frac{E}{I} \qquad \text{(Equation 5-6)}$$

$$P = I \times E \qquad \text{(Equation 5-7)}$$

$$L_{TOTAL} = L_1 + L_2 + L_3 + \ldots + L_n \qquad \text{(Equation 5-8)}$$

$$L_{TOTAL} = \cfrac{1}{\cfrac{1}{L_1} + \cfrac{1}{L_2} + \cfrac{1}{L_3} + \ldots + \cfrac{1}{L_n}} \qquad \text{(Equation 5-9)}$$

$$L_{TOTAL} = \frac{L_1 \times L_2}{L_1 + L_2} \qquad \text{(Equation 5-10)}$$

$$C_{TOTAL} = C_1 + C_2 + C_3 + ... + C_n \qquad \text{(Equation 5-11)}$$

$$C_{TOTAL} = \frac{1}{\frac{1}{C_1} + \frac{1}{C_2} + \frac{1}{C_3} + ... + \frac{1}{C_n}} \qquad \text{(Equation 5-12)}$$

$$C_{TOTAL} = \frac{C_1 \times C_2}{C_1 + C_2} \qquad \text{(Equation 5-13)}$$

$$Bw = 2 \times (D + M) \qquad \text{(Equation 8-1)}$$

$$\lambda = \frac{c}{f} \qquad \text{(Equation 9-1)}$$

$$\lambda = \frac{3 \times 10^2}{f(MHz)} = \frac{300}{f(MHz)} \qquad \text{(Equation 9-2)}$$

$$\frac{\lambda}{2} = \frac{150}{f(MHz)} \qquad \text{(Equation 9-3)}$$

$$L \ (ft) = \frac{468}{f(MHz)} \qquad \text{(Equation 9-4)}$$

$$L_{director} = L_{driven} \times 0.95 \qquad \text{(Equation 9-5)}$$

$$L_{reflector} = L_{driven} \times 1.05 \qquad \text{(Equation 9-6)}$$

Circumference of driven element:
$$C_{driven\ element}(ft) = \frac{1005}{f(MHz)} \qquad \text{(Equation 9-7)}$$

Circumference of director element:
$$C_{director}(ft) = \frac{975}{f(MHz)} \qquad \text{(Equation 9-8)}$$

Circumference of reflector element:

$$C_{reflector} \text{ (ft)} = \frac{1030}{f(MHz)}$$ (Equation 9-9)

$$SWR = \frac{E_{max}}{E_{min}}$$ (Equation 9-10)

$$SWR = \frac{Z_0}{R} \text{ or } SWR = \frac{R}{Z_0}$$ (Equation 9-11)

Appendix C

Glossary of Key Words

Alternating current (ac)—Electrical current that flows first in one direction in a wire and then in the other direction. The applied voltage changes polarity and causes the current to change direction. This direction reversal continues at a rate that depends on the frequency of the ac.

Amplifier—A device usually employing electron tubes or transistors to increase the voltage, current, or power of a signal. The amplifying device may use a small signal to control voltage and/or current from an external supply. A larger replica of the small input signal appears at the device output.

Attenuate—To reduce in amplitude.

Audio-frequency shift keying (AFSK)—A method of transmitting radio-teletype information by switching between two audio tones fed into an FM transmitter microphone input. This is the RTTY mode most often used on VHF and UHF.

Backscatter—A small amount of signal that is reflected from the earth's surface after traveling through the ionosphere. The reflected signals may go back into the ionosphere along several paths and be refracted to earth again. Backscatter can help provide communications into a station's skip zone.

Balanced line—A symmetrical feed line with two conductors having equal but opposite voltages. Neither conductor is at ground potential.

Balun—A transformer used between a BALanced and an UNbalanced system. Used for feeding a balanced antenna with an unbalanced feed line.

Band plan—An agreement for operating within a certain portion of the radio spectrum. Band plans set aside certain frequencies for each different mode of amateur operation, such as CW, SSB, FM, repeaters and simplex.

Band-pass filter—A circuit that allows signals to go through it only if they are within a certain range of frequencies. It attenuates signals above and below this range.

Bandwidth—The frequency range (measured in hertz) over which a signal is stronger than some specified amount below the peak signal level. For example, if a certain signal is at least half as strong as the peak power level over a range of ±3 kHz, the signal has a 3-dB bandwidth of 6 kHz.

Beat-frequency oscillator (BFO)—An oscillator that provides a signal to the product detector. In the product detector, the BFO signal and the IF signal are mixed to produce an audio signal.

Breakdown voltage—The voltage at which an insulating material will conduct current.

Capacitor—An electrical component composed of two or more conductive plates separated by an insulating material. A capacitor stores energy in an electrostatic field.

Carbon-composition resistor—An electronic component designed to limit current in a circuit; made from ground carbon mixed with clay.

Carbon-film resistor—A resistor made by putting a gaseous carbon deposit on a round ceramic form.

Ceramic capacitor—An electronic component composed of two or more conductive plates separated by a ceramic insulating material.

Clipping—Occurs when the peaks of a voice waveform are cut off in a transmitter, usually because of overmodulation. Also called **flattopping.**

Coaxial cable—Feed line with a central conductor surrounded by plastic, foam or gaseous insulation. In turn it is covered by a shielding conductor. The entire cable is usually covered with vinyl insulation.

Coil—A conductor wound into a series of loops.

Color code—A system where numerical values are assigned to various colors. Colored stripes are painted on the body of resistors and sometimes other components to show their value.

Core—The material in the center of a coil. The material used for the core affects the inductance value of the coil.

Critical angle—If radio waves leave an antenna at an angle greater than the critical angle for that frequency, they will pass through the ionosphere instead of returning to earth.

Critical frequency—The highest frequency at which a vertically incident radio wave will return from the ionosphere. Above the critical frequency, radio signals pass through the ionosphere instead of returning to the earth.

Cubical quad antenna—An antenna built with its elements in the shape of four-sided loops.

Current—A flow of electrons in an electrical circuit.

Cutoff frequency—In a high-pass, low-pass, or band pass filter, the cutoff frequency is the frequency at which the filter output is reduced to 1/2 of the power available at the filter input.

D layer—The lowest layer of the ionosphere. The D layer contributes very little to short-wave radio propagation. It acts mainly to absorb energy from radio waves as they pass through it. This absorption has a significant effect on signals below about 7.5 MHz during daylight.

Delta loop antenna—A variation of the cubical quad antenna, with triangular elements.

Detector—The stage in a receiver in which the modulation (voice or other information) is recovered from the RF signal.

Deviation ratio—The ratio between the maximum change in RF-carrier frequency and the highest modulating frequency used in an FM transmitter.

Dielectric—The insulating material used between the plates in a capacitor.

Dielectric constant—A number used to indicate the relative "merit" of an insulating material. Air is given a value of 1, and all other materials are related to air.

Director—A parasitic element in "front" of the driven element in a multi-element antenna.

Direct current (dc)—Electrical current that flows in one direction only.

Direct wave—A radio wave traveling in a straight line between the transmitting antenna and the receiving antenna.

Direct-conversion receiver—A receiver that converts an RF signal directly to an audio signal with one mixing stage.

Driven element—The element connected directly to the feed line in a multi-element antenna.

Duct—A radio waveguide formed when a temperature inversion traps radio waves within a restricted layer of the atmosphere.

Dummy load (dummy antenna)—A resistor that acts as a load for a transmitter, dissipating the output power without radiating a signal. A dummy load is used when testing transmitters.

E layer—The second lowest ionospheric layer, the E layer exists only during the day, and under certain conditions may refract radio waves enough to return them to earth.

Electric field—An invisible force of nature. An electric field exists in a region of space if an electrically charged object placed in the region is subjected to an electrical force.

Electrolytic capacitor—A polarized capacitor formed by using thin foil electrodes and chemical-soaked paper.

Electromotive force (EMF)—The force or pressure that pushes a current through a circuit.

Emission designator—A symbol made up of two letters and a number, used to describe a radio signal. A3E is the designator for double-sideband, full-carrier, amplitude-modulated telephony.

F layer—A combination of the two highest ionospheric layers, the F1 and F2 layers. The F layer refracts radio waves and returns them to earth. The height of the F layer varies greatly depending on the time of day, season of the year and amount of sunspot activity.

Farad—The basic unit of capacitance.

Feed line—The wire or cable used to connect an antenna to the transmitter and receiver.

Field-effect transistor volt-ohm milliammeter (FET VOM)—A multiple-range meter used to measure voltage, current and resistance. The meter circuit uses an FET amplifier to provide a high input impedance for more accurate readings.

Filter—A circuit that will allow some signals to pass through it but will greatly reduce the strength of others.

Fixed resistor—A resistor with a fixed nonadjustable value of resistance.

Flattopping—See **Clipping**.

Frequency Coordinator—A volunteer who keeps records of repeater input, output and control frequencies.

Frequency deviation—The amount the carrier frequency in an FM transmitter changes as it is modulated.

Frequency modulation—The process of varying the frequency of an RF carrier in response to the instantaneous changes in a modulating signal. The signal that modulates the carrier frequency may be audio, video, digital data or some other kind of information.

Frequency-shift keying (FSK)—A method of transmitting radioteletype information by switching an RF carrier between two separate frequencies. This is the RTTY mode most often used on the HF amateur bands.

Front-to-back ratio—The energy radiated from the front of a directional antenna divided by the energy radiated from the back of the antenna.

Gain—An increase in the effective power radiated by an antenna in a certain desired direction. Also, an increase in received signal strength from a certain direction. This is at the expense of power radiated in, or signal strength received from, other directions.

Gamma match—A method of matching coaxial feed line to the driven element of a multielement array.

Geometric or **True horizon**—The most distant point one can see by line of sight.

Guided propagation—Radio propagation by means of ducts.

Half-wavelength dipole antenna—A fundamental antenna one-half wavelength long at the desired operating frequency. It is connected to the feed line at the center. This is a popular amateur antenna.

Henry—The basic unit of inductance.

High-pass filter—A filter that allows signals above the cutoff frequency to pass through. It attenuates signals below the cutoff frequency.

Horizontally polarized wave—An electromagnetic wave with its electric lines of force parallel to the ground.

Induced EMF—A voltage produced by a change in magnetic lines of force around a conductor. When a magnetic field is formed by current in the conductor, the induced voltage always opposes changes in that current.

Inductor—An electrical component usually composed of a coil of wire wound on a central core. An inductor stores energy in a magnetic field.

Intermediate frequency (IF)—The output frequency of a mixing stage in a superheterodyne receiver. The subsequent stages in the receiver are tuned for maximum efficiency at the IF.

Ionosphere—A region in the atmosphere about 30 to 260 miles above the earth. The ionosphere is made up of charged particles, or ions.

Low-pass filter—A filter that allows signals below the cutoff frequency to pass through and attenuates signals above the cutoff frequency.

Major lobe—The shape or pattern of field strength that points in the direction of maximum radiated power from an antenna.

Marker generator—An RF signal generator that produces signals at known frequency intervals. The marker generator can be used to calibrate receiver and transmitter frequency readouts.

Maximum usable frequency (MUF)—The highest frequency that allows a radio wave to reach a desired destination.

MAYDAY—From the French "m'aider" (help me), MAYDAY is used when calling for emergency assistance in voice modes.

Metal-film resistor—A resistor formed by depositing a thin layer of resistive-metal alloy on a cylindrical ceramic form.

Mica capacitor—A capacitor formed by alternating layers of metal foil with thin sheets of insulating mica.

Mixer—A circuit used to combine two or more audio- or radio-frequency signals to produce a different output frequency.

Modulate—To vary the amplitude, frequency, or phase of a radio-frequency signal.

Modulation—The process of varying some characteristic (amplitude, frequency or phase) of an RF carrier for the purpose of conveying information.

Modulation index—The ratio between the maximum carrier frequency deviation and the frequency of the modulating signal at a given instant in an FM transmitter.

Monitor Oscilloscope—A test instrument connected to an amateur transmitter and used to observe the shape of the transmitted-signal waveform.

Multimeter—An electronic test instrument used to make basic measurements of current and voltage in a circuit. This term is used to describe all meters capable of making different types of measurements, such as the VOM, VTVM and FET VOM.

Mutual coupling—When coils display mutual coupling, a current flowing in one coil will induce a voltage in the other. The magnetic flux of one coil passes through the windings of the other.

Offset—The difference between a repeater's input and output frequencies. On 2 meters, for example, the offset is either plus 600 kilohertz (kHz) or minus 600 kHz from the receive frequency.

Ohm—The basic unit of resistance.

Ohm's Law—A basic law of electronics, it gives a relationship between voltage, resistance and current ($E = IR$).

Oscillator—A circuit built by adding positive feedback to an amplifier. It produces an alternating current signal with no input except the dc operating voltages.

Paper capacitor—A capacitor formed by sandwiching paper between thin foil plates, and rolling the entire unit into a cylinder.

Parallel circuit—An electrical circuit in which the electrons follow more than one path in going from the negative supply terminal to the positive terminal.

Parallel-conductor feed line—Feed line constructed of two wires held a constant distance apart. They may be encased in plastic or constructed with insulating spacers placed at intervals along the line.

Parasitic element—Part of a directive antenna that derives energy from mutual coupling with the driven element. Parasitic elements are not connected directly to the feed line.

Phase—If you consider a point on a waveform, phase is the angular difference between that point and any other point on the same, or another, waveform.

Phase modulation—Varying the phase of an RF carrier in response to the instantaneous changes in the modulating signal.

Plastic-film capacitor—A capacitor formed by sandwiching thin sheets of Mylar™ or polystyrene between thin foil plates, and rolling the entire unit into a cylinder.

Polarization—The orientation of the electric lines of force in a radio wave, with respect to the surface of the earth.

Potentiometer—A resistor whose resistance can be varied continuously over a range of values.

Propagation—The means by which radio waves travel from one place to another.

Quarter-wavelength vertical antenna—An antenna constructed of a quarter-wavelength-long radiating element placed perpendicular to the earth.

Radio-path horizon—The point where radio waves are returned by tropospheric bending. The radio-path horizon is 15 percent farther away than the geometric horizon.

Reactance—The property of an inductor or capacitor (measured in ohms) that impedes current in an ac circuit without converting power to heat.

Reactance modulator—A device capable of modulating an ac signal by varying the reactance of a circuit in response to the modulating signal. (The modulating signal may be voice, data, video, or some other kind depending on what type of information is being transmitted.) The circuit capacitance or inductance changes in response to an audio input signal.

Reflected wave—A radio wave whose direction is changed when it bounces off some object in its path.

Reflectometer—A test instrument used to indicate standing wave ratio (SWR) by measuring the forward power (power from the transmitter) and reflected power (power returned from the antenna system).

Reflector—A parasitic element placed "behind" the driven element in a directive antenna.

Refract—To bend. Electromagnetic energy is refracted when it passes through a boundary between different types of material. Light is refracted as it travels from air into water or from water into air.

Resistance—The ability to oppose an electrical current.

Resistor—Any material that opposes a current in an electrical circuit. An electronic component specifically designed to oppose current.

Rotor—The movable plates in a variable capacitor.

RST System—The system used by amateurs for giving signal reports. "R" stands for readability, "S" for strength and "T" for tone. See Table 2-1.

S meter—A meter in a receiver that shows the relative strength of a received signal.

Selectivity—A measure of how well a receiver can separate a desired signal from other signals on a nearby frequency.

Sensitivity—The ability of a receiver to detect weak signals.

Series circuit—An electrical circuit in which all the electrons must flow through every part of the circuit. There is only one path for the electrons to follow.

Sidebands—The sum or difference frequencies generated when an RF carrier is mixed with an audio signal.

Signal generator—A test instrument that produces a stable low-level radio-frequency signal. The signal can be set to a specific frequency and used to troubleshoot RF equipment.

Simplex—A term normally used in relation to VHF and UHF operation, simplex operation means you are receiving and transmitting on the same frequency.

Skip zone—A region between the farthest reach of ground-wave communications and the closest range of skip propagation.

Sky waves—Radio waves that travel from an antenna upward to the ionosphere, where they either pass through the ionosphere into space or are refracted back to earth.

Solar flux index—A measure of solar activity. The solar flux index is a measure of the radio noise on 2800 MHz.

$\overline{\text{SOS}}$—A Morse code call for emergency assistance.

Space wave—A radio wave arriving at the receiving antenna made up of a direct wave and one or more reflected waves.

Splatter—The term used to describe a very wide bandwidth signal, usually caused by overmodulation of a sideband transmitter. Splatter causes interference to adjacent signals.

Stability—A measure of how well a receiver or transmitter will remain on frequency without drifting.

Standing wave ratio—the ratio of maximum voltage to minimum voltage along a feed line. Also the ratio of antenna impedance to feed-line impedance when the antenna is a purely resistive load.

Stator—The stationary plates in a variable capacitor.

Sunspots—Dark blotches that appear on the surface of the sun.

Superheterodyne receiver—A receiver that converts RF signals to an intermediate frequency before detection.

Temperature inversion—A condition in the atmosphere in which a region of cool air is trapped beneath warmer air.

Toroidal inductor—A coil wound on a donut-shaped ferrite or powdered-iron form.

Transmission line—See Feed line.

Troposphere—The region in the earth's atmosphere just above the surface of the earth and below the ionosphere.

Tropospheric bending—When radio waves are bent in the troposphere, they return to earth approximately 15 percent farther away than the geometric horizon.

True or **Geometric horizon**—The most distant point one can see by line of sight.

Twin lead—See **Parallel-conductor feed line**.

Unbalanced line—Feed line with one conductor at ground potential, such as coaxial cable.

Vacuum-tube voltmeter (VTVM)—A multimeter using a vacuum tube in its input circuit. The VTVM does not load a circuit as much as a VOM, and can be used to measure small voltages in low-impedance circuits. A VTVM generally requires 120 V ac power.

Varactor diode—A component whose capacitance varies as the reverse-bias voltage changes.

Variable capacitor—A capacitor that can have its value changed within a certain range.

Variable resistor—A resistor whose value can be adjusted over a certain range.

Variable-frequency oscillator (VFO)—An oscillator used in receivers and transmitters. The frequency is set by a tuned circuit using capacitors and inductors. The frequency can be changed by adjusting the components in the tuned circuit.

Vertically polarized wave—A radio wave that has its electric lines of force perpendicular to the surface of the earth.

Virtual height—The height that radio waves appear to be reflected from when they are returned to earth by refraction in the ionosphere.

Voltage—The force or pressure (EMF) that pushes a current through a circuit.

Volt-ohm-milliammeter (VOM)—A test instrument used to measure voltage, resistance and current. A VOM usually has a RANGE switch so the meter can be used to measure a wide range of inputs.

Wattmeter—A test instrument used to measure the power output (in watts) of a transmitter.

Wire-wound resistor—A resistor made by winding a length of wire on an insulating form.

Yagi antenna—A directive antenna made with a half-wavelength driven element. It has one or more parasitic elements arranged in the same horizontal plane.

Index

Notes

Notes

Notes

Notes